# STRUCTURE FROM MOTION IN THE GEOSCIENCES

## New Analytical Methods in Earth and Environmental Science

Introducing New Analytical Methods in Earth and Environmental Science, a new series providing accessible introductions to important new techniques, lab and field protocols, suggestions for data handling and interpretation, and useful case studies.

This series represents an invaluable and trusted source of information for researchers, advanced students, and applied earth scientists wishing to familiarise themselves with emerging techniques in their field.

All titles in this series are available in a variety of full-colour, searchable e-book formats. Titles are also available in an enhanced e-book edition which may include additional features such as DOI linking and high-resolution graphics and video.

*Ground-Penetrating Radar for Geoarchaeology*
by Lawrence B. Conyers

*Rock Magnetic Cyclostratigraphy*
by Kenneth P. Kodama and Linda A. Hinnov

*Techniques for Virtual Palaeontology*
by Mark Sutton, Imran Rahman, and Russell Garwood

# STRUCTURE FROM MOTION IN THE GEOSCIENCES

JONATHAN L. CARRIVICK
MARK W. SMITH
DUNCAN J. QUINCEY

WILEY Blackwell

This edition first published 2016 © 2016 by John Wiley & Sons, Ltd

*Registered Office*
John Wiley & Sons, Ltd, The Atrium, Southern Gate, Chichester, West Sussex, PO19 8SQ, UK

*Editorial Offices*
9600 Garsington Road, Oxford, OX4 2DQ, UK
The Atrium, Southern Gate, Chichester, West Sussex, PO19 8SQ, UK
1606 Golden Aspen Drive, Suites 103 and 104, Ames, Iowa 50010, USA

For details of our global editorial offices, for customer services and for information about how to apply for permission to reuse the copyright material in this book please see our website at www.wiley.com/wiley-blackwell.



*Library of Congress Cataloging-in-Publication data applied for*

ISBN: 9781118895849

A catalogue record for this book is available from the British Library.

Wiley also publishes its books in a variety of electronic formats. Some content that appears in print may not be available in electronic books.

Cover image: Illustration of SfM -MVS workflow which progresses from multiple overlapping images to point cloud generation and usually then to differenced elevation model: in this case for detecting and analysing erosion and deposition in an alpine river © Jonathan Carrivick

Set in 10/12.5pt Minion by SPi Global, Pondicherry, India

1 2016

# Contents

# Abbreviations

| | |
|---|---|
| **ALS** | airborne laser scanning |
| **DEM** | digital elevation model |
| **dGPS** | differential Global Positioning System |
| **DoD** | DEM of difference |
| **DSLR** | digital single-lens reflex camera |
| **EXIF** | exchangeable image file format |
| **Gb** | gigabyte |
| **LiDAR** | light detection and ranging |
| **MAE** | mean absolute error |
| **Mb** | megabyte |
| **NRSfM** | non-rigid Structure from Motion |
| **PMVS** | patch-based multi-view stereo |
| **RAM** | random access memory |
| **RANSAC** | random sample consensus |
| **RGB** | red–green–blue |
| **RMSE** | root-mean-square error |
| **SfM-MVS** | Structure from Motion–Multi-View Stereo |
| **Tb** | terrabyte |
| **TLS** | terrestrial laser scanning |
| **TS** | total station |
| **UAV** | unmanned aerial vehicle |

# About the Companion Website

This book is accompanied by a companion website:

www.wiley.com/go/carrivick/structuremotiongeosciences

The website includes the following:

- Videos
- Figures and tables from the book for downloading
- Interactive figures

# 1 Introduction to Structure from Motion in the Geosciences

Abstract

Structure from Motion (SfM) is a topographic survey technique that has emerged from advances in computer vision and traditional photogrammetry. It can produce high-quality, dense, three-dimensional (3D) point clouds of a landform for minimal financial cost. As a topographic survey technique, SfM has only been applied to the geosciences relatively recently. Its flexibility, particularly in terms of the range of scales it can be applied to, makes it well suited to a field as diverse as the geosciences. This book is designed to act as a primer for scientists and environmental consultants working within the geosciences who are interested in using SfM or are seeking to understand more about the technique and its limitations. The early chapters consider SfM as a method within the context of other digital surveying techniques, and detail the SfM workflow, from both theoretical and practical standpoints. Later chapters focus on data quality and how to measure it using independent validation before looking in depth at the range of studies that have used SfM for geoscience applications to date. This book concludes with an outward look towards where the greatest areas of potential development are for SfM, summarising the main outstanding areas of research.

Keywords

geosciences; Structure from Motion; multi view stereo; GIS; landform

## 1.1 The Geosciences and Related Disciplines

Geoscience is a term that encompasses many disciplines of research and industry, particularly environmental consultancy. It is an umbrella term for climate, water and biogeochemical cycles, and planetary tectonics, which are the three basic processes that shape the Earth's surface. These are complex natural

*Structure from Motion in the Geosciences*, First Edition. Jonathan L. Carrivick, Mark W. Smith, and Duncan J. Quincey.

Companion Website: www.wiley.com/go/carrivick/structuremotiongeosciences

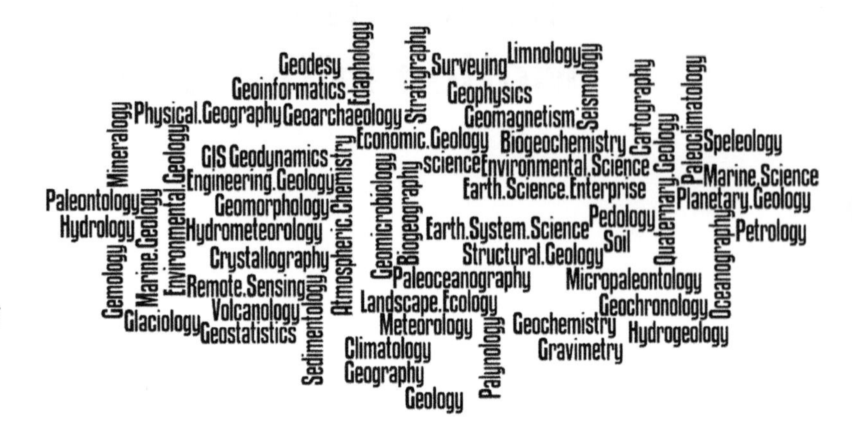


**Figure 1.1** A word cloud of geoscience sub-disciplines. The font size of each word does not indicate anything.

systems in space and time. For example, process responses and interactions occur on spatial scales spanning hundreds of kilometres to microns, such as river catchments and abrasion marks on fluvially-transported grains, respectively. Process responses and interactions occur on timescales ranging from picoseconds for chemical reactions to millions of years for plate tectonics and biological evolution, respectively.

Whether academic or applied, whether large scale or small scale, the geosciences seek to understand the forces and factors that shape our world and the environments in which we live. Reasons for requiring understanding of these forces and factors span many remits: exploitation for the hydrocarbon and renewable energy sectors, managing natural hazards, managing a resource-consuming and dynamic society, mitigating effects of climate change, and academic interest and enquiry, for example.

In seeking ever-refined understanding for application to real-world problems, the geosciences now transcend "traditional" earth science disciplines (Fig. 1.1). The multidisciplinary nature of the geosciences is partly due to it having become particularly adept at pursuing interactions between the biological, chemical, and physical sciences. Analysis across these traditional boundaries is critical to understanding systems in an integrated and holistic manner.

Furthermore, the geosciences are now established as being notable for embracing emerging and novel technologies and innovations. Indeed the revolution brought about by spatial analysis software such as geographical information systems (GIS) has been argued as a new paradigm in the discipline. Many technologies in the geosciences have been adapted from the military, from the petroleum industries, and more recently from computer science.

No matter what particular specialism to which they affiliate themselves, many geoscience disciplines will generally recognise three key tasks:

1 Recognition of spatial patterns
2 Documentation of transient landforms
3 Linking processes to products

Here it is important to note that in this book we use the term **landform** independent of any scale; the term as used in this book is considered to encompass landscape, terrain, feature, surface, and texture, for example.

A common requirement for each of these three key tasks is for the geosciences to have topographic information. The primary function of topographic information in digital format is to quantify landform variability and more specifically three-dimensional (3D) structure.

Topographic information can be used to identify landforms and landform properties. Landforms can include natural and artificial features and thus form part of the description of a specific place. When landforms are observed to change, the processes causing those changes are often inferred conceptually, and perhaps also tested by numerical models. New methods of acquiring topographic data with a fine spatial resolution are to be welcomed because they expose greater detail about landform morphology. They also provide an opportunity to match the scale of topographic data with the spatiotemporal scale of the landform or processes under investigation.

On the basis of the multidisciplinary nature of the geosciences and of the widespread academic and applied need for topographic survey data, we consider that this book, which will focus specifically on one specific method for generating topographic data, has relevance for all the geosciences (Fig. 1.1) and for related disciplines. Related disciplines requiring topographic information include architecture, archaeology, civil engineering and subdisciplines associated with built structures, objects, and artefacts, and biology and medicine where concerns range from vegetation to anatomical surveys.

## 1.2 Aim and Scope of this Book

The aim of this book is to describe an emerging survey method and workflow that is better established in related disciplines such as archaeology (e.g. De Reu et al. 2013) and cultural heritage (Koutsoudis et al. 2014) and is now finding widespread uptake in the geosciences, namely, "Structure from Motion" (SfM). This book is designed to act as a primer for geoscientists who are interested in using SfM or are seeking to understand more about the technique and its limitations.

This book is designed to appeal to students, professional academics, and industry practitioners, particularly environmental consultants. Whilst existing texts dealing with SfM are often heavily mathematical, originating commonly from computer vision literature, this book is designed to be fully accessible by an interested geoscience audience that may not necessarily be fully conversant in complex mathematical operations involved in SfM. Thus, the workflow of SfM is described in a predominantly qualitative manner, and the reader is referred elsewhere for further technical details. Important terms are in bold at first use, a list of abbreviations is provided, and emphases of particularly important properties are in italics. This book is designed to balance the conceptual discussion of application, theory, and technical details of analytical methods. Thereby this book serves as a synoptic reference to both inform and educate. In educating, we emphasise the discussion and development of a critical understanding of the application of SfM in the geosciences to date. In terms of informing, we build on this critical understanding to stimulate ideas for carefully considered future developments of the SfM workflow by the geosciences.

## 1.3 The Time and the Place

This book is timely and of immediate relevance because of (i) the emergence of an affordable, user-friendly software; (ii) rapid developments in unmanned aerial vehicles (UAVs) or drones and other potential SfM survey platforms; and (iii) a dearth of textbooks on SfM in the geosciences. Notwithstanding that, of course, the pace of technological change in hardware and software is incredibly rapid, and for that reason the forward-looking chapters of this book do not dwell on specific hypothetical applications but rather on major themes and concepts.

At present, the use of SfM can only really be evaluated in academic literature since technical and industry reports tend not to be listed on public databases. A search in the academic publications database *Web of Knowledge* for *Structure from Motion* (made in April 2015) delivered approximately 1000 records since the early 1980s (Fig. 1.2). Computer science was the category with the most counts of that phrase. Engineering was ranked 2nd and geosciences was ranked 9th. Notably, the geosciences have only started producing publications incorporating SfM in the past decade (Fig. 1.2).

The impact of SfM is arguably going to be greater than that associated with the advent of airborne laser scanning (ALS) or airborne light detection and ranging (LiDAR), not least because SfM workflows democratise data collection and the development of fine-resolution 3D models at all scales of landscapes, landforms, surfaces, and textures. Moreover, to produce such advanced data products, very little input data are required: as little as a photograph set from an uncalibrated, compact (and therefore often cheap) camera. In a similar vein to airborne LiDAR surveys and terrestrial laser scanning (TLS) 15 and 10 years ago, respectively, the past couple of years have seen a raft of sessions at major international conferences describing work using SfM. This book places these developments in context, outlines the analytical framework and key issues, and presents 10 detailed case studies contributed by SfM practitioners.

This book fills a niche where there is a current dearth of textbooks on SfM for the geosciences. Although existing photogrammetry-orientated and

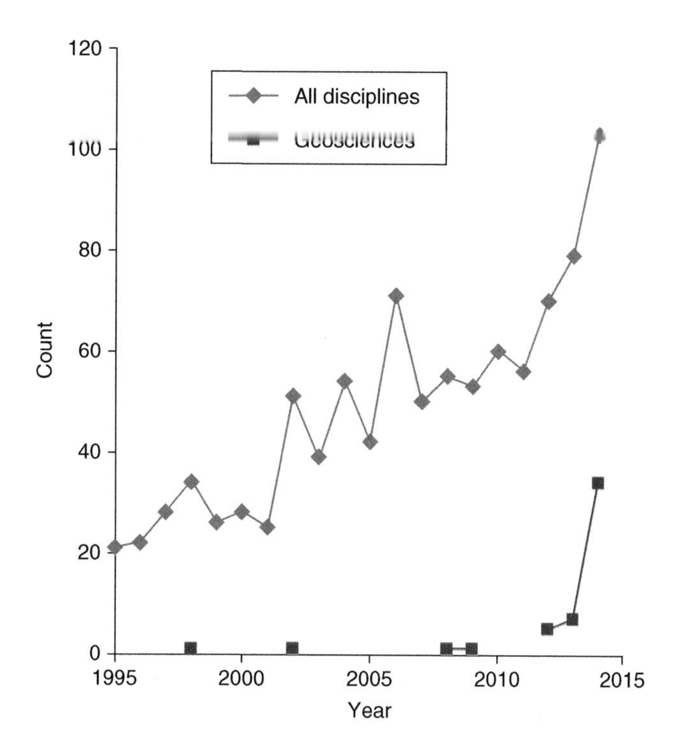


**Figure 1.2** Count of citations to "Structure from Motion" in the academic literature.

computer vision-oriented texts will describe in depth many of the algorithms and procedures that SfM utilises, albeit in a modified or improved form to handle the input of dozens or hundreds of images, these texts are often extremely technical and may be largely inaccessible to the "average" geoscientist. The Wiley-published book series "New Analytical Methods in Earth and Environmental Science" has two other titles in that series that may be complementary to this book: *Techniques for Virtual Palaeontology* (Sutton et al. 2014) and a proposed *Digital Outcrop Modelling*. Both of these certainly fall within the geosciences domain and illustrate that geoscience usage of SfM is not just about terrain models.

## 1.4 What Is Structure from Motion?

Structure from Motion (also known as Structure-and-Motion) has developed since the 1980s into a valuable tool for generating 3D models from 2D imagery, not least with the development of software with graphical user interfaces (GUIs). Full details of the SfM workflow are provided in Chapter 3 but are summarised briefly here. In contrast to traditional photogrammetry, SfM uses algorithms to identify matching features in a collection of overlapping digital images and calculates camera location and orientation from the differential positions of multiple matched features. Based on these calculations overlapping imagery can be used to reconstruct a "sparse" or "coarse" 3D point cloud model of the photographed object or surface or scene. This 3D model from the SfM method is usually refined to a much finer resolution using Multi-View Stereo (MVS) methods, thereby completing the full SfM-MVS workflow. Whilst there is prevailing practice in the geosciences literature to abbreviate this workflow simply to **SfM**, we herein use the SfM-MVS acronym to give clarity and precision to the workflow, and thus to champion rigorous practice in reflecting the different aspects of the full workflow. More details on the distinction between SfM and MVS are provided in Chapter 3.

In brief, the exciting and attractive properties of SfM-MVS are that it is cheap in both hardware and software requirements, is fast in comparison to other digital surveying in the field, and is a workflow that is virtually independent of spatial scale. Furthermore, SfM-MVS can produce a spatial density/resolution of survey points and 3D point accuracy that in some circumstances is comparable to that from modern terrestrial laser scanners. However, SfM-MVS is still in its infancy, especially in the geosciences, and as is explored in this book, more technical research needs to be done to understand the quality of data produced.

## 1.5 Structure of this Book

This book broadly comprises a critical commentary on the present usage of SfM-MVS alongside other digital surveying methods in the geosciences, an appraisal of the SfM-MVS workflow and data products, and a consideration of future developments of SfM-MVS that the geosciences could exploit.

All of the main chapters have boxes detailing **case studies**, and these have been contributed either by researchers invited for their particularly interesting and novel use of SfM-MVS in the geosciences, or else case studies have been provided by the authors of this book as examples of innovative use of SfM-MVS. All of the main chapters have hyperlinked text to relevant websites and online material. Some figures contain links to an interactive or animated version of that figure, or to an equivalent example online. All chapters have suggestions for further reading.

Specifically, Chapter 2 outlines the place of SfM-MVS in the geosciences in the context of other existing digital surveying methods via qualitative and graphical comparisons of advantages and disadvantages. The properties of the resulting data are compared quantitatively with those produced by other methods.

Chapter 3 presents a thorough outline of the SfM-MVS workflow. Technical details of each step in the SfM-MVS workflow are outlined to provide the reader with a greater understanding of how exactly the SfM-MVS method works. As such, Chapter 3 provides a useful stand-alone reference for geoscientists who wish to know more about SfM-MVS, without needing to delve into the computer vision literature. Steps in the described workflow include detection of corresponding features in multiple images, reconstruction of camera position, and derivation of a sparse point cloud, MVS to derive a dense point cloud, scaling and georeferencing, and optimisation.

Chapter 4 presents practical and logistical details that a practitioner would need to know before undertaking an SfM-MVS survey. As a flexible survey method, there are many choices to be made when designing a survey, with advantages and disadvantages to each choice. These decisions include, and thus this chapter compares, different cameras, platforms, processing software, point cloud viewers, and methods of constructing 2.5d terrain models (or digital elevation models (DEMs)) from point clouds, specifically filtering, decimation, surface interpolation/reconstruction, and surface rendering.

Whilst Chapter 2 focuses on a qualitative comparison with other digital survey methods, Chapter 5 presents a more quantitative analysis by summarising existing validation studies of SfM-MVS for the first time. Many SfM-MVS validation studies have emerged in the geosciences in recent years; however, each deploys a slightly different method, and there is little standardisation of such validation. This reflects that the use of SfM-MVS in the geosciences is still in its infancy. Synthesis of existing data provides useful indication of the achievable data quality given any particular SfM-MVS survey design and thus the potential of SfM-MVS in the geosciences.

Chapter 6 summarises the way in which SfM-MVS is being used in the geosciences today. A wide variety of applications have been found for this technique over a wide range of spatial and temporal scales and in many different environments. Current applications are reviewed, and key science questions being asked are examined. Several case studies have been chosen to emphasise the breadth and depth of contrasting applications of SfM-MVS in the geosciences.

Chapter 7 offers suggestions as to the potential development of SfM-MVS for the geosciences not only in terms of hardware and software but also in terms of whole avenues and themes of study yet to be explored. Chapter 7 discusses major project types that have yet to be exploited by the geosciences; namely, automatic detection, augmented reality, real-time mapping, and non-rigid SfM-MVS. The deliberate aim of Chapter 7 is not to look for incremental developments or simply for different applications. Rather, it is more ambitious, firstly examining developments in SfM-MVS in other disciplines and secondly using this information to suggest where the geosciences should look to develop itself.

Chapter 8 summarises the main outstanding areas of research identified in this book. Gaps in knowledge and potential future directions are indicated. Given the rapid development of SfM-MVS in the geosciences and the potential revolution it could bring to our discipline, these are exciting times indeed.

## References

De Reu, J., Plets, G., Verhoeven, G. et al. (2013) Towards a three-dimensional cost-effective registration of the archaeological heritage. *Journal of Archaeological Science*, **40** (**2**), 1108–1121.

Koutsoudis, A., Vidmar, B., Ioannakis, G., Arnaoutoglou, F., Pavlidis, G. & Chamzas, C. (2014) Multi-image 3D reconstruction data evaluation. *Journal of Cultural Heritage*, **15** (**1**), 73–79.

Sutton, M., Rahman, I. & Garwood, R. (2014) *Techniques for Virtual Palaeontology*, p. 208. John Wiley & Sons, Inc., Hoboken, NJ.

# The Place of Structure from Motion

## *A New Paradigm in Topographic Surveying?*


**Abstract**

Three-dimensional (3D) spatial data on the form of the earth's surface are of paramount importance to geoscientists seeking to document the shape of landforms and to understand land-forming and natural hazard processes. Commonly, spatiotemporal information on landform changes is required. Therefore, advances in digital survey and sensor technology that address spatial and temporal constraints have created new opportunities to investigate the structure and dynamics of landform systems. Structure from Motion (SfM) represents the latest and a very significant advance in digital surveying. This chapter places SfM in the context of existing techniques, namely, total stations (TS), differential Global Positioning Systems (dGPS), photogrammetry, airborne laser scanning (ALS), and terrestrial laser scanning (TLS). TS points are surveyor determined and thus whilst slow to gather, they can be carefully selected to produce the most efficient representation of topography and are unlikely to include artefacts. GPS survey equipment can be cumbersome to transport by hand, it but does deliver spatial data in real-world coordinates and can gather thousands of points per hour. ALS is very expensive and only achieved on a campaign basis but does offer broader spatial coverage than ground-based methods, typically at 1–2 m point spacing. TLS can offer point spacing at millimetre scale and can include truly 3D information such as within cliff undercuts resulting in multiple surface levels at a single 2D coordinate. In comparison, Structure from Motion (SfM) is very cheap and fast, can offer truly 3D information, and with careful use of ground control points (GCPs) can rival other digital survey methods for spatial accuracy.



**Keywords**

topography; topographic survey; photogrammetry; global positioning system; LiDAR; laser scanning; total station; digital elevation model; point cloud

*Structure from Motion in the Geosciences*, First Edition. Jonathan L. Carrivick, Mark W. Smith, and Duncan J. Quincey.



Companion Website: www.wiley.com/go/carrivick/structuremotiongeosciences

## 2.1 Introduction

This chapter considers Structure from Motion–Multi-View Stereo (SfM-MVS) in the context of the fundamental requirement of the geosciences to acquire topographic data. Geologists, earth scientists, geographers, environmental scientists, environmental engineers, and those in related disciplines have always required field data on the shape or form of natural surfaces. Commonly, information on changes to landforms is also required, so multi-temporal surveys are needed to detect such changes. Advances in digital survey and sensor technology have created new opportunities to investigate the structure and dynamics of earth surface systems through the development of high-quality digital elevation models (DEMs) and techniques of differencing these models for the detection of geomorphological change. The same digital topographic data are of utmost importance to studies concerned with landform "roughness," such as surface energy balance and surface hydrodynamics.

SfM-MVS represents the latest significant advance in digital surveying and is being used increasingly as a crucial component of the geoscientist's toolkit (Westoby et al. 2012; Carrivick et al. 2013a). However, there exists a wide range of alternative surveying methods of which SfM-MVS is just one. Topographic data for the geosciences can be focussed at different spatial scales (Fig. 2.1). With consideration of the scale of survey, this chapter places SfM-MVS in the context of these existing techniques. The chapter herein is qualitative; Chapter 5 presents a quantitative validation of SfM-MVS. Approaches to acquiring digital topography by the geosciences can be categorised as either direct or indirect. Direct approaches require contact by the surveyor with the landform of interest. Indirect approaches permit measurement of a landform whilst the surveyor remains remote from that landform.

This chapter first outlines the basic properties of different survey platforms and digital sensors, indicating the survey design and methods that accompany each one. In doing so, it clarifies the types of geoscience application to which SfM-MVS could be applied to maximum effect. For instance, the choice of survey technique will depend on the intended usage of the data and on constraints such as time and money and required expertise for hardware operation and data processing. The second part of this chapter discusses the advantages and challenges of SfM-MVS in the context of digital surveys in the geosciences.

## 2.2 Direct Topographic Surveying

### *2.2.1 Total Stations*

Digital topographic surveying with total stations (TS) or electronic distance measurement devices is most advantageously used where high precision of a few (<100) single points is required and in relatively enclosed natural

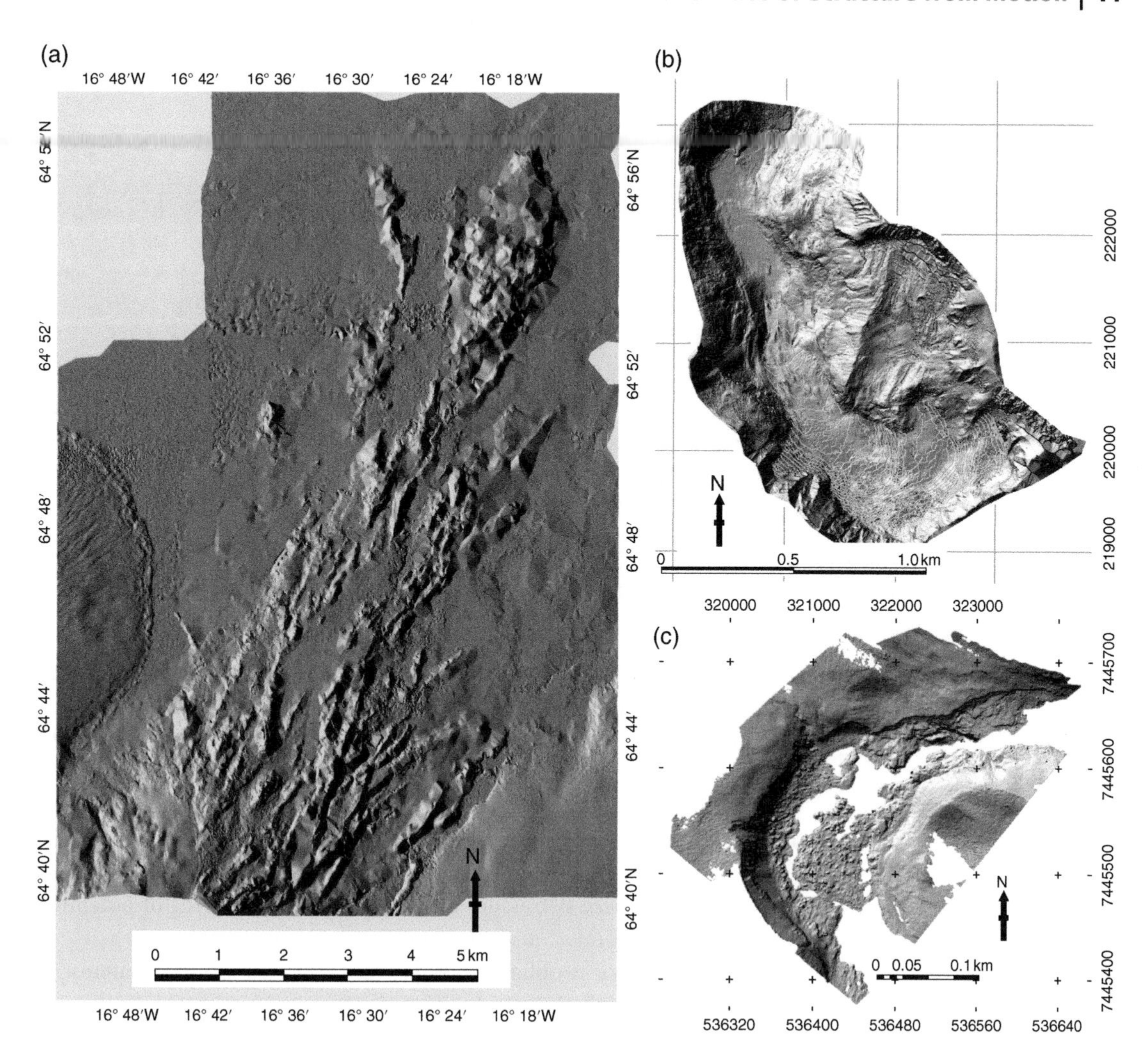


**Figure 2.1** Topographic data visualised as a hillshaded elevation model and spanning scales of (a) landscape, (b) valley, and (c) reach scales. All of these might be termed as fine resolution, given the spatial extent of the data set; panel (a) depicts a 5 m grid resolution DEM of the Kverkfjöll–Kverkfjallarani area, central Iceland, as derived from stereo-photogrammetry of aerial photographs (Carrivick & Twigg 2005), panel (b) depicts a 2 m grid resolution DEM of the Ödenwinkelkees alpine catchment, central Austria, as derived from airborne laser scanning (Carrivick et al. 2013b), and panel (c) depicts a 0.5 m grid resolution DEM of a bedrock river gorge on the northern flank of Russell Glacier, West Greenland, as derived from Structure from Motion.

topography such as canyons where sky view is limited, and thus where Global Positioning System (GPS) usage may not be possible. TS survey points can be focussed on breaklines that best delineate breaks of slope and thus can produce the most accurate representation of landform topography. Furthermore, because TS points are chosen individually and expertly, they are unlikely to include artefacts, which can be a problem with data from remote survey methods (see Section 2.3). The expert judgment of where and when to survey points also makes TS surveys, with static dGPS surveys, the

(a) (b)

**Figure 2.2** Topographic surveying with a total station (TS): a prism reflector target (a) and a tripod-mounted TS (b) in use in southern Spain.

most efficient of all digital surveying approaches (Lane et al. 1994; Brasington et al. 2003; Wheaton et al. 2010).

Standard TS equipment requires two operators: one at the station and the other at the target point of interest (Fig. 2.2). This makes them expensive to operate and also limits targets to those directly accessible in person. Therefore, robotic and reflectorless TS (e.g. Keim et al. 1999; Fuller et al. 2003; Tsai et al. 2012; Brown & Pasternack 2014) have been developed in part to address these two issues and with the added benefit of potentially speeding up the point acquisition rate.

When a surveyor is obliged to physically visit each point of interest, an element of subjectivity is introduced to the survey design because (i) spatial coverage (survey area extent and three-dimensional or 3D point density) must be balanced against time and money and accessibility and (ii) the geometry of the points surveyed fundamentally affects the quality of the interpolated surface if they are converted to a digital elevation model. Therefore, grid-based sampling, cross-section-based surveys, or topographically stratified survey designs can be implemented (e.g. Vallé & Pasternack 2006). TS measurements are restricted to being in a local coordinate system, unless the survey is "tied" or "back-sighted" to a point with "known" real-world coordinates.

### 2.2.2 *Differential GPS*

Real-world coordinates can be provided instantaneously by Global Navigation Satellite Systems (GNSS), usually the America-based GPS but increasingly also Russia's space-based GLObal NAvigation Satellite System (GLONASS)

(a)

(b)

**Figure 2.3** Topographic surveying with a differential Global Positioning System: a "temporary" base receiver mounted on a tripod with an external radio transmitter (a), and the corresponding rover unit set up for a static occupation using a mini tripod and a radio receiver (b), both in Tarfala Valley, Arctic Sweden (see the website for the videos).

and the Europe-based Galileo system. GPS has been used for surveying and mapping since the 1990s, and initially it was mainly employed to provide control points for traditional triangulation-based surveying techniques. However, since the mid-1990s, GPS has been used as a surveying method in its own right. The simplest application of GNSS involves recording the location of points of interest, for example, individual boulders, where a single reading may suffice. With larger or more complex landforms a surveyor must select a series of points to define the landform geometry.

Acquisition of differential GPS (dGPS) data requires the surveyor to visit each point of interest with a *rover* receiver. This rover receiver can be mounted on the surveyor's back or on a survey pole, depending on the balance between point accuracy and speed of survey chosen. Positioning of the rover can be calculated in real time relative to a "temporary" *base* station receiver, which is set up on a tripod over a known point (Fig. 2.3a) and this real-time function is facilitated via radio link; hence, a modem and a radio antenna are required at both the base and the rover. Alternatively, rover data can be post-processed relative to the temporary base station, or relative to a continuous or permanent base station such as those part of national and international geodetic system networks.

In "continuous" mode, several thousands of points per hour can be obtained. One type of continuous survey is the *real-time kinematic* mode that requires a direct radio or mobile telephone modem link between the base and rover receiver but does have the benefit of providing the final accuracy to the surveyor in the field at the point of interest. If dGPS survey data are interpolated to create a DEM, changes in volume can be calculated using raster subtraction between surfaces (e.g. Fig. 2.4). In *static* mode, where multiple observations are logged and averaged per point, GNSS systems can

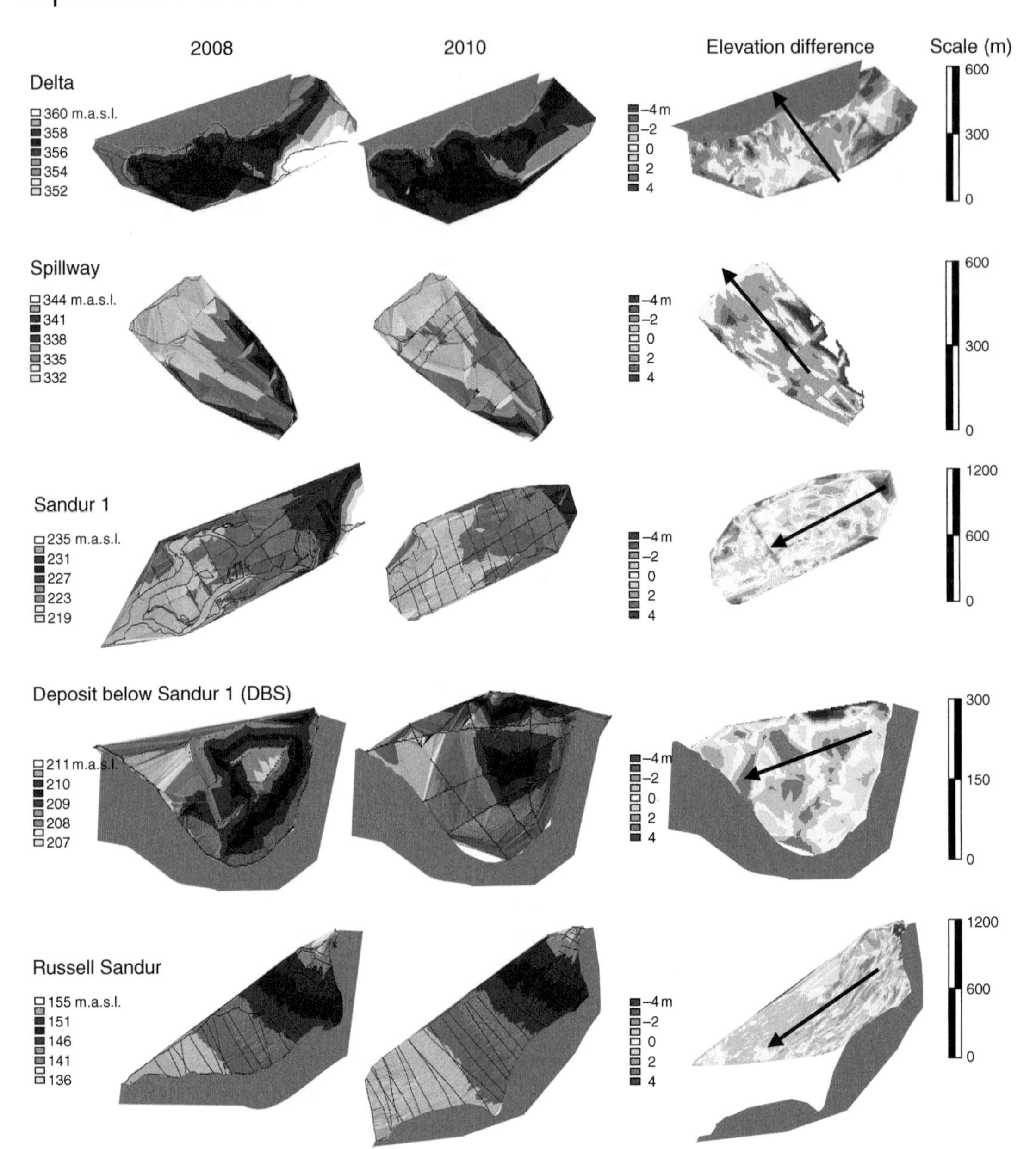


**Figure 2.4** Landform changes as determined by differencing of an interpolated elevation surface through each of 2008 and 2010 differential Global Positioning System points (black dots) at five sites along a glacial meltwater-fed river at Russell Glacier, western Greenland. Note the blue masked area is water. The black arrows indicate the direction of water flow. Source: From Carrivick et al. (2013c).

be used to acquire up to several tens of points per hour. This survey mode is of particular use if the GNSS-derived 3D point data are being used either as input to other survey methods as ground control points (GCPs) or to validate the accuracy of other survey methods.

**Table 2.1** Accuracy and productivity of different methods of conducting a dGPS survey.

| Method | Foot | Foot | Wheel | Vehicle (1 antenna) | Vehicle (2 antenna) |
|---|---|---|---|---|---|
| dGPS survey mode | Static | Continuous | Continuous | Continuous | Continuous |
| Accuracy (m) | <0.005 | ~0.05 | <0.1 | <0.1 | <0.05 |
| Speed (points per second) | <0.05 | 1 | 10 | 100 | 100 |

Adapted from Young (2013).
Note that the values given are indicative of some of the highest achievable rates on open terrain. Surveys on more complex terrain, or where there is some tree cover or in proximity to buildings, are likely to take far longer to complete.

dGPS survey data are commonly imported to geographic information systems (GIS), where linear points can be converted to lines and survey points to DEMs via interpolation routines (e.g. inverse distance weighting, nearest neighbour analysis, kriging), where the type of interpolation routine applied depends on the point coverage and type of surface of interest. This GIS processing of GNSS data makes comparisons with other data sets possible. Repeat GNSS surveys of the same landforms can be compared, making dGPS a useful tool for measuring surface changes; for example flood-induced geomorphological changes (Fig. 2.4), or for measuring glacier surface velocities "on the ground" in contrast to via remote sensing.

dGPS data accuracy is dependent on the number and the geometry of satellites used to compute a point and also on the equipment set-up and survey mode used (Table 2.1 after Young 2013). Surveyor expertise can also be an issue in the accuracy of DEMs created from dGPS points since survey points are chosen judgementally, as in TS surveys (Bangen et al. 2014). dGPS point accuracy can be sub-centimetre by increasing point occupation times, that is, "static" survey mode rather than continuous or kinematic mode (Table 2.2), and by post-processing using data from the International Geodetic System network. This sub-centimetre accuracy of GNSS data compares favourably with that from other digital survey methods, being similar to that of terrestrial laser scanning (TLS) (e.g. Brasington et al. 2000; Casas et al. 2006; Hugenholtz et al. 2013), and more accurate than airborne laser scanning (ALS), which has a highest achievable accuracy of approximately 0.1–0.2 m, for example, Sallenger et al. (2003). Additionally, dGPS accuracy is not significantly reduced by environmental factors, although maintaining (an uninterrupted) radio link between a temporary base and a rover can be awkward in high winds, high humidity, or near highly reflective surfaces due to blowing dust, moisture, and heat haze, respectively. Proximity to trees and buildings creates big problems with maintaining satellite reception and real-time radio lock.

**Table 2.2** Advantages and disadvantages of different dGPS survey methods.

| Method | Advantages | Disadvantages |
|---|---|---|
| Foot (static) | Very high accuracy | Very low productivity |
| | No modifications to equipment required | |
| | Point density can be varied | Accuracy and productivity may be reduced in high winds |
| | Suitable for all terrain | |
| Foot (continuous) | No modifications to equipment required | Error introduced as antenna not always vertical (especially on slopes) |
| | Points taken automatically | |
| | Suitable for all terrain | |
| Wheel | No expensive modifications required | Error introduced as antenna not always vertical |
| | Points taken automatically | Wheel may sink into very soft sediments |
| | Suitable for most terrain | |
| Vehicle (1 antenna) | Very high productivity | Expensive modifications required |
| | | Unsuitable for some terrains (e.g. fragile soils and plants and steep slopes) |
| | | Vehicle may sink into soft sediments |
| | Points taken automatically | Requires vehicle access |
| | | Error introduced as antenna not always vertical (especially on uneven terrain) |
| Vehicle (2 antennae) | Very high productivity | Expensive modifications required |
| | | Unsuitable for some terrains (e.g. fragile soils and plants and steep slopes) |
| | | Vehicle may sink into soft sediments |
| | High accuracy | Requires vehicle access |
| | | Error introduced as antenna not always vertical (especially on uneven terrain) |

After Young (2013) with permission.

## 2.3 Remote Digital Surveying

Automation of digital surveying has increased point acquisition rate, increased spatial coverage (survey area extent and 3D point density) and removed the judgment of sample point selection. This spatially seamless surveying has been termed "hyperscale" surveying by Brasington et al. (2012). Detection of changes between surveys has increased in temporal resolution. Perhaps most crucially remote digital surveying removes the need for a surveyor to physically visit the target point of interest and so offers opportunities for surveying inaccessible landforms. Remote digital surveying includes digital photogrammetry, laser scanning, and also SfM-MVS, all of which initially produce 3D point clouds. It must be emphasised that each of these remote methods requires precise "ground control points (GCPs)" distributed across the study area for georeferencing of the survey data in real-world coordinates, which are commonly obtained via either TS or dGPS (Section 4.4).

### 2.3.1 Photogrammetry

Applications of photogrammetry in the geosciences have usually obtained photographs via survey-grade analogue cameras, often mounted onboard piloted aircraft. Traditional photogrammetric methods require precise knowledge of the 3D location and pose of the camera and the precise 3D location of a series of control points in the scene. Using the former, triangulation can be used to reconstruct scene geometry, whilst in the latter control points are manually identified in the input photographs and a process called "resectioning" or "camera pose estimation" is used to determine the camera position. This resectioning process and indeed the whole photogrammetry workflow is often labour intensive and lengthy. Aircraft enable the best combination of spatial coverage and ground resolution, which is a function of flying height, to be obtained. Digital photogrammetry is well used in the geosciences across several spatial scales (e.g. Chandler 1999; Lane et al. 2000, 2010; Baltsavias et al. 2001; Chandler et al. 2002; Mora et al. 2003; Carbonneau et al. 2003; Bitelli et al. 2004; Lim et al. 2005; Sturzenegger & Stead 2009; Fischer et al. 2011; Ribeiro et al. 2013; Staines et al. 2014) and from a variety of ground-based and aerial platforms. These studies predominantly consider rapidly changing surfaces such as braided rivers, landslides, coastal cliffs, and glaciers.

This widespread usage in the geosciences of photogrammetry has been enhanced with the development of methods allowing for the accurate calibration of non-metric cameras and the increasingly reliable automation of the photogrammetric process (e.g. Chandler 1999; Chandler et al. 2002; Carbonneau et al. 2004). In particular, the development of semi-automated digital triangulation and image-based landform extraction algorithms, of free and open-source software, and the rapid technological advancement and ever-decreasing cost of high-end desktop computers have significantly enhanced the quality of photogrammetry-derived DEMs. In a significant departure from using vertical (overhead) photographs, Chandler (1999), and others since, has implemented a series of rotation matrices in photogrammetry software in order to process oblique, ground-based images thereby enhancing the flexibility of that workflow. Nonetheless, image acquisition for photogrammetry requires careful consideration in finding a suitable vantage point, whether that be ground-based or an aerial platform, and these are discussed in detail in Chapter 4. Furthermore, in terms of methodology, whilst photogrammetry has prevailed in the geosciences, it has a major shortcoming in requiring images with known distortion properties, expert understanding, and expert practice. For example, photogrammetry requires near-parallel stereo pairs of images with approximately 60% overlap, accurate measurement of the camera position and accurate camera calibration.

We note that DEMs typically of tens of metres in resolution are commonly obtained from stereo satellite imagery, such as ASTER, which is also the source of the GDEM-2, a global DEM with 30 m pixel spacing. DEMs of a few metres in horizontal resolution from high-resolution satellite images have recently been produced using specially designed algorithms, for processing stereo pairs of DigitalGlobe imagery (e.g. Noh & Howat 2015).

## 2.3.2 Laser Scanning

Laser scanning has created a step change in the spatiotemporal coverage that can be incorporated in topographic surveys and also in the speed of point acquisition, with millions of points per hour being readily achievable with most systems. However, this voluminous data collection has created problems in processing and analysing the data and raises questions over approaches to processing.

Laser scanners emit a laser pulse and record the time it takes for that pulse to return to the scanner. They do this many thousands or tens of thousands of times per second. Lasers travel at a constant speed and because the direction in which the laser was emitted is known, the distance from the scanner of any reflecting surface and also the remote coordinates of the point of reflection are known. This process is generally termed light detection and ranging, or "LiDAR." In this manner "point clouds" comprising millions of points are compiled in just a few minutes. Laser scanners can be mounted on aircraft or on tripods on the ground (Fig. 2.5).

### 2.3.2.1 Airborne Laser Scanning

Laser scanners can comprise systems designed to be mounted on aircraft (airborne laser scanning: ALS; Fig. 2.5a) and integrated with inertial measurement units for positioning and correction or aircraft pitch, yaw, and roll. There has been a proliferation of geoscience-based researchers using ALS for glacier monitoring, river evolution, and hillslope mass movements,

(a)

(b)

**Figure 2.5** Examples of an airborne LiDAR system: an auxiliary pod fastened beneath the wing of a lightweight Dimona aircraft (a) and a terrestrial laser scanner mounted on a tripod (b). Both examples are during use in the Ödenwinkelkees catchment, central Austria (see the website for the videos).

for example (e.g. Charlton et al. 2003; Hilldale & Raff 2008; Knoll & Kerschner 2010; Bertoldi et al. 2011; Joerg et al. 2012; Lin et al. 2013; Razak et al. 2013; Carrivick et al. 2009a,b, 2010, 2013b).

The accuracy of ALS data largely depends on the dGPS and inertial measurement unit systems (Hodgson & Bresnahan 2004) and in complex natural environments can result in large vertical offsets in surfaces (Heritage & Milan 2009). Other factors affecting ALS data accuracy involve both internal technical parameters and settings of the ALS system and external factors such as object reflectance, atmospheric conditions, landform slope, or density of vegetation cover. Generally, the vertical accuracy of ALS data is between 0.1 and 0.15 m and the horizontal accuracy between 0.1 and 0.5 m (Baltsavias 1999; table 1 in Gallay 2013). Perhaps most importantly for the geosciences, vertical accuracy of ALS data deteriorates with increasing landform slope angle (Hodgson & Bresnahan 2004). ALS data users are also frequently confronted with systematic errors in digital terrain model/digital surface model (DTM/DSM) data such as artefacts and mismatch of flight strips and distortions in the rendering of data (Gallay 2013).

Several empirical studies have revealed vertical accuracies of ALS DTMs between 0.08 and 0.33 m root mean square error (RMSE), which were subject to parameters of the platform and environmental conditions (Gallay 2013). Bangen et al. (2014) compared an ALS-derived DEM with that from a TS and a dGPS and found it to have a typical grid cell elevation difference of 0.5 m (Fig. 2.6).

#### 2.3.2.2 Terrestrial Laser Scanning

Terrestrial laser scanners comprise units that are mounted *in situ* on conventional survey tripods (Fig. 2.5b) and thus permit rapid (typically tens of minutes) re-survey of dynamic landforms (e.g. Rosser et al. 2005; Jones et al. 2007; Milan et al. 2007; Hodge et al. 2009; Notebaert et al. 2009; McCoy et al. 2010; Brasington et al. 2012; Carrivick et al. 2013b; Williams et al. 2014; Abellán et al. 2014) as well as definition of forest tree biomass (e.g. Kankare et al. 2013; Srinivasan et al. 2014), soil hydraulic roughness (Smith et al. 2011), snow pack ablation (e.g. Egli et al. 2012), aeolian saltation (Nield et al. 2011) and lava flow evolution (Nelson et al. 2011), for example. A review of TLS methods and data processing has recently been produced by Smith (2015).

Terrestrial laser scanners are optimised for precision at a given range because the laser beam spreads (i.e. has a larger foot print) with increasing distance from the instrument. Scanners (for outdoor mobile use) exist for surveys of landforms from tens of metres distance (e.g. Leica C10, Leica HDS7000 and Faro Focus) to hundreds (e.g. Riegl VZ1000 and Maptek iSite 8810) and thousands of metres (Riegl VZ400 and VZ8000 and Optech ILRIS-IS). Largely as a consequence of the laser required to achieve this range, the speed, point accuracy, and other properties of the data from the TLS can vary markedly. Point acquisition rate or speed can be from thousands to hundreds of thousands of points per second.

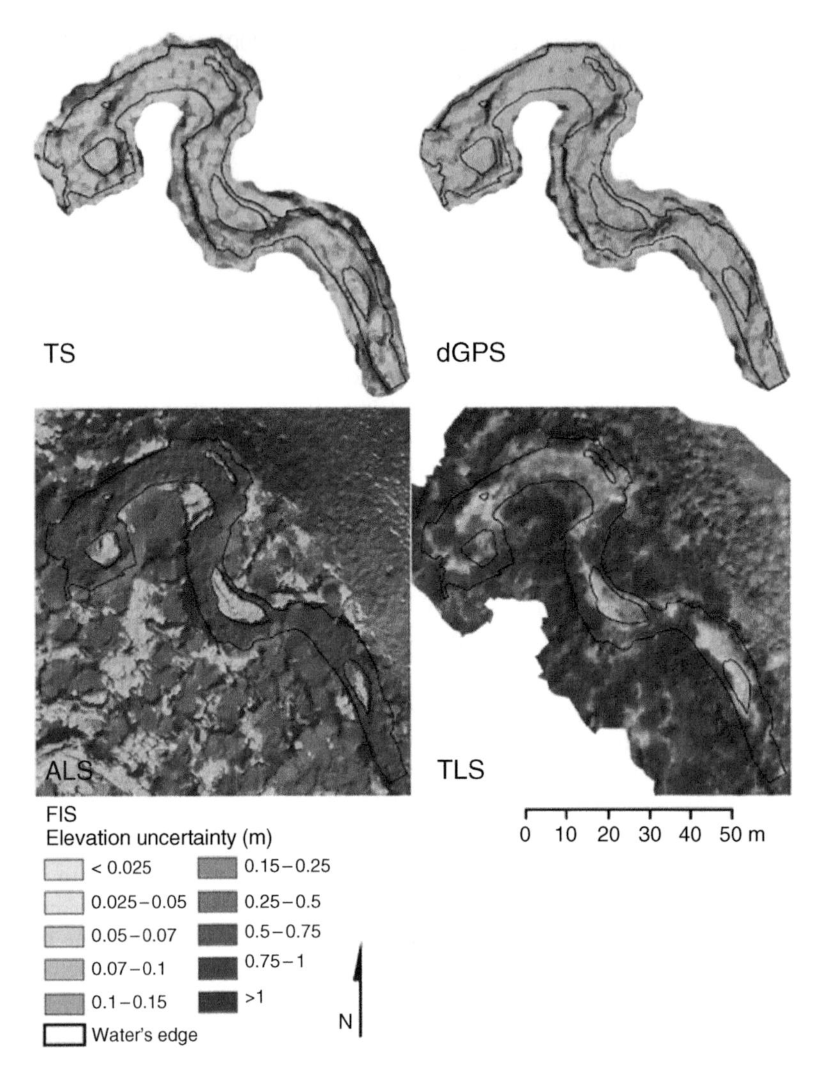


**Figure 2.6** Example of estimated elevation uncertainties compared for total station (TS), differential Global Positioning System (dGPS) used in real-time kinematic mode, airborne laser scanning (ALS), and terrestrial laser scanning (TLS), at Bear Valley Creek. All were derived using separate Fuzzy inference system (FIS) models (Wheaton et al. 2010), except the TLS-derived DEM which is a roughness model derived from the detrended standard deviation of the point cloud. Source: With permission after Bangen et al. (2014).

#### 2.3.2.3 Advantages of Laser Scanning

Laser scanners offer a number of benefits over other survey methods besides the obvious merits of spatial extent/coverage and speed (Alho et al. 2009). As an active remote sensing technique, ALS mapping is not dependent on the time of day or land cover. In comparison with photogrammetry, sampling landform heights with ALS is independent of the diversity of image texture. Therefore, ALS enables reliable 3D landform reconstruction even for areas with snow cover or sand.

Some airborne (e.g. Riegl VQ and LMS series) and terrestrial laser scanners (e.g. Riegl VZ series) have "full waveform" capability. Full waveform systems digitise the complete returned signal, rather than a few selected maxima, and through decomposition algorithms deliver better (higher resolution) vertical characterisation of the target. Furthermore, by analysing the pulse width and amplitude of individual echoes, it is possible to provide information on the backscattering properties of the target. These benefits offer significant potential for a range of geoscience applications, including forestry, landform analysis, and classification. A review of full waveform laser scanning has been given by Mallet and Bretar (2009).

A key advantage of ALS over other digital survey methods is that some laser pulses are able to penetrate through sparse vegetation enabling both vegetation height and "bare-earth" elevation to be determined simultaneously. Some scanners have the ability to scan in the green wavelength which permits through water scanning of bathymetry, at least in relatively shallow, still, and clear water (e.g. via ALS: Irish & Lillycrop 1999 and via TLS: Smith et al. 2012). The measurement of water depth relies on the differential timing of laser pulses reflected from the water surface (infrared laser) and the underwater surface (green laser) to determine the water depth at the point where the laser pulses strike the water surface (Cavalli & Tarolli 2011).

Some terrestrial laser scanners record a red–green–blue (RGB) value to add colour to the point cloud or have an integral camera. Virtually all airborne and terrestrial scanners record the intensity of the laser received back at the scanner, which can give insight to the material properties of the remote landform. Unlike ALS systems, which tend to "look downwards," terrestrial laser scanners can produce full 3D point clouds due to being positioned at an oblique angle to the surface of interest.

#### 2.3.2.4 Disadvantages of Laser Scanning

Laser scanning is not a perfect solution to topographic surveying. Perhaps most obviously the cost of aircraft deployment is a major limiting factor for geoscience research (Slatton et al. 2007). Furthermore, the amount of data acquired with ALS for a given area can be excessive for processing with conventional computer, especially with full waveform mapping when data volumes can increase by 50–200 times in comparison to single return ALS techniques (Mallet & Bretar 2009). The high level of detail captured in ALS data can be regarded as a constraint in cases where processes or phenomena are to be observed or modelled on a larger scale (Wood 2009).

Some TLS systems can obtain laser returns from landforms at approximately 8 km distance, but since the laser has a large "footprint" or beam width at this distance, it is generally scanners with less than 1 km range that have precision necessary for many geoscience applications. Multiple scan positions are likely to be necessary, not only to avoid blind spots behind obstacles but also to gain the spatial coverage (survey areas extent and 3D point density) required. The high hardware costs (~£30,000–£120,000) and labour-intensive acquisition can limit the extent and frequency of surveys; a TLS system including accessories weighs approximately 40 kg and targets for georeferencing, which must be chosen carefully so as to be visible from multiple scan positions, that must be visited for survey with TS or dGPS. For large and remote surveys, therefore, preparation in survey design is important and may benefit from GIS-based analysis of line-of-sight, sky view, and buffer distances from each possible scan position and target position, as achieved for a TLS survey covering 9 $km^2$ by Carrivick et al. (2015) in the Tarfala Valley, Arctic Sweden.

Uncertainty in TLS-derived DEMs can be greater where signals are scattered by wet surfaces (Charlton et al. 2003; Milan et al. 2007) or as a result of vegetation returns misinterpreted as ground shots where vegetation or woody debris prevented the laser signal from reaching the ground (e.g. Heritage & Milan 2009). Several of the larger magnitude TLS-derived DEM elevation uncertainties identified by Bangen et al. (2014) were difficult to interpret (Fig. 2.6), but they are thought to have resulted from (i) the scanner collecting returns from within the wetted channel that were not properly corrected for refraction and appeared below the true surface elevation or (ii) spurious artefacts and nodes removed during the triangulated irregular network (TIN) editing process but resulting in elevation that was artificially lowered.

There are often issues with occlusion (hidden surfaces) and the need to filter out non-surface points such as "mixed pixels," which result from multiple surfaces of different ranges falling within the laser beam footprint and the resultant averaging out (or mixing) of both returns to yield a non-surface point. Point cloud–based analyses of laser scanner data are very much in development, but some open-source software does exist and should provide ample opportunity for novel analyses (Brasington et al. 2012; Rychkov et al. 2012; Kreylos et al. 2013). Thus the default action at present for geoscientists is to grid the topographic data, which often loses information or introduces interpolation errors. Analyses of full-waveform data from either airborne or terrestrial laser scanners (e.g. Hakala et al. 2012) have yet to be exploited fully by geoscientists; to date being focussed on forest canopy and stem structure analyses (e.g. Wagner et al. 2008; Liang et al. 2012; Yang et al. 2013) and more general vegetation properties (Mallet & Bretar 2009; Lindberg et al. 2012; Hollaus et al. 2014).

### 2.3.3 *Comparison of Digital Survey Methods*

SfM-MVS is a workflow that utilises multiple (overlapping) images of a landform from a photographic camera from multiple viewpoints to reconstruct 3D object or landform geometry. In contrast to traditional photogrammetry, scene geometry, camera positions, and orientation are retrieved simultaneously by SfM-MVS and without requirement for 3D camera position and pose or 3D position of GCPs to be known first. However, SfM-MVS does require GCPs for scale, and sets of photographs with a high degree of overlap and which capture the full three dimensionality of the landform of interest, as viewed from a range of surrounding positions and aspects; that is, as the name implies, imagery derived from a non-static sensor.

Whilst the mode of operation and spatio-temporal coverage of topographic data obtained via digital survey method is an important factor for consideration (Table 2.3), to demonstrate fully the pros and cons of each

**Table 2.3** Summary of properties and pros and cons of different digital survey methods.

| Survey equipment and workflow | Typical spatial extent (km)/ typical spatial resolution (pt.m²) | Possible data acquisition rate (points per hour) | Possible 3D point accuracy (m) | Advantages | Disadvantages |
|---|---|---|---|---|---|
| TS | 0.1–1.0/0.1–5.0 | Hundreds | <0.001 | Low cost<br>Accurate | Line of sight required<br>Low productivity<br>Accuracy decreases with distance from base |
| dGPS | 2.4–1.0/0.1–5.0 | Thousands | 0.005 | High accuracy<br>Range of methods have been developed to suit different surveying requirements<br>Line of sight not required | High cost<br>Some methods have low productivity<br>Lock on 6+ satellites required |
| Photogrammetry | 5.0–50.0/ 0.5–10.0 | Tens of thousands | 0.5 | High productivity<br>Once set up, no operator required<br>Continuous information can be captured | Low resolution<br>Equipment must be left in position for long periods of time (depending on survey) and may be vandalised or damaged<br>Does not work in fog, mist, etc. |
| ALS | 5.0–100.0/ 0.2–10.0 | Millions | 0.2 | High productivity<br>Can be used during the night<br>Airborne LiDAR can survey areas that are difficult to access<br>Not affected by vegetation cover | Very high cost<br>Resolution may be insufficient to measure small changes<br>Systematic errors on some landforms |
| TLS | 0.01–5.0/ 100–10,000 | Millions | 0.05 | High accuracy | Unable to capture all aspects of complex topographies (depending on equipment positioning) |
| SfM | 0.01–1.0/ 1–10,000 | Millions | 0.01–0.2 | Cheap<br>Fast<br>Method is independent of spatial scale | Reproducibility?<br>Reliability? |

Advantages and disadvantages are adapted with permission from Young (2013). Extent and resolution values are from figure 12 in Bangen et al. (2014).

method, other factors such as capital expenditure costs, maintenance and running costs, labour (number of persons and time) costs, and last but by no means least processing of that data must be evaluated. Inevitably there is a trade-off between mobility enabled by aircraft or motorised vehicles, which incur "rental" costs, versus manually moving by person which is free but takes a lot of time and incurs subsistence costs. Aircraft costs depend on flight time, which is not just a function of the area to be surveyed but also of the distance from the nearest airport.

If digital survey methods are ranked by capital expenditure, where that listed first is greatest, then the order is ALS, TLS, dGPS, TS, photogrammetry, and lastly (cheapest) SfM-MVS (Table 2.3; Fig. 2.7). The ranking is the same for the maintenance and running costs, which include insurance, hardware servicing and software licensing, and support fees. If the methods are ranked by labour costs, the order is ALS, photogrammetry, TLS, TS, dGPS, and lastly (cheapest) SfM-MVS (Table 2.3; Fig. 2.7). This is because at minimum (i) ALS requires a pilot and an instrument operator to be airborne and largely automated post-processing of the data; (ii) photogrammetry requires a lot of (relatively manual) post-processing; and (iii) TLS usually requires two people in the field (both initially to move the heavy equipment and then one to operate the scanner and often one to collect target positions). dGPS and TS (if robotic or reflectorless) and SfM-MVS can be operated by one person in the field at a minimum.

#### 2.3.3.1 Advantages of SfM-MVS

More specifically, the main attraction of SfM-MVS in comparison to other digital survey methods is that it is cheap. Only a camera and a desktop computer are required. Software is freely available. In contrast, a TLS costs tens of thousands of GBP and requires licensed software and frequent professional servicing. Furthermore, SfM-MVS surveys are relatively easy to undertake. Hand-held cameras present no issues of field portability, unlike terrestrial laser scanners which weigh several tens of kilograms. SfM-MVS can be applied with images obtained from a camera mounted on one of many types of platform, and can be processed with a number of commercial or open-source software or code.

A further notable difference of SfM-MVS in comparison to other digital survey methods is that SfM-MVS produces fully 3D data, as otherwise only possible with TLS (Section 2.3.2.2). SfM-MVS–derived point clouds and textured surfaces are inherently multi-dimensional (x, y, z, point orientation, colour, texture), and relatively easily transformed into orthophotographs and into DEMs (Bemis et al. 2014). See examples of a boulder, of an experimental gravel-bed flume, and of an entire river reach (see the companion website for the interactive figures).

SfM-MVS is a workflow that can be applied to spatial scales ranging from $10^{-2}$ to $10^{6}\,m^2$ (Smith & Vericat 2015), and that workflow remains virtually identical regardless of the spatial and temporal scales under consideration, though the achievable survey quality is dependent on survey range. With

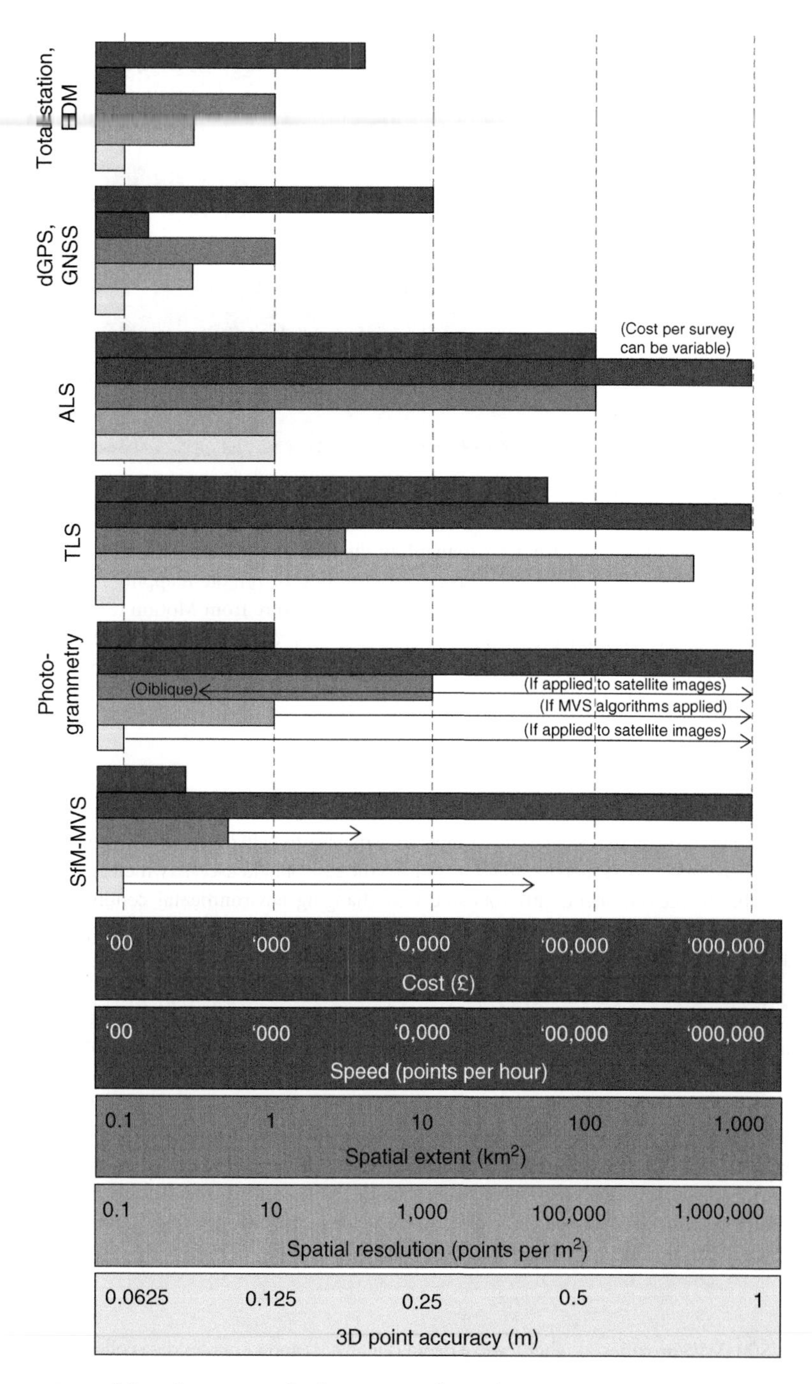


**Figure 2.7** Comparison of digital survey methods with regard to financial cost, maximum possible speed, spatial coverage, resolution, and accuracy. Note that the photogrammetry and SfM-MVS values are completely dependent on survey range. Bars derived from information in Brasington et al. (2000), Young (2013), Gallay (2013), and Bangen et al. (2014).

## Box 2.1 Case study: Structure from Motion versus the Kinect: A comparison of river field measurements at the $10^{-2}$–$10^{2}$ m scales

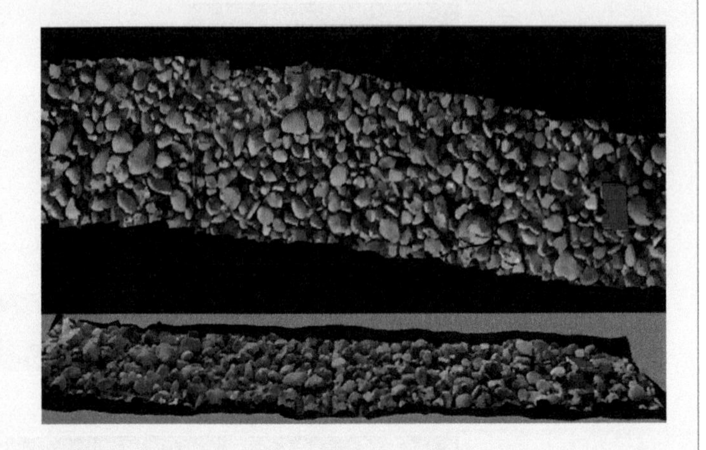

Mark A. Fonstad, University of Oregon
James T. Dietrich, Dartmouth College

### Background and context

At the smaller spatial scales of fluvial field analysis, measurements historically made *in situ* are beginning to be supplemented, or even replaced by remote sensing methods by agencies and researchers. This is particularly true in the case of topographic and particle size measurement. In the field, the scales of *in situ* observation usually range from millimetres up to hundreds of metres. Two recent approaches for remote mapping of river environments at the scales of historical *in situ* observations are (i) camera-based Structure from Motion (SfM-MVS) and (ii) active patterned-light measurement with devices such as the Kinect. Even if only carried by hand, these two approaches can produce topographic data sets over three to four orders of magnitude of spatial scale. Which approach currently is most useful?

### Method

Previous studies have demonstrated that both SfM-MVS and the Kinect are precise and accurate over *in situ* field measurement scales; we instead turn to alternate comparative metrics to help determine which tools might be best for our river measurement tasks. These metrics might include (i) the ease of field use, (ii) which general environments are or are not amenable to measurement, (iii) robustness to changing environmental conditions, (iv) ease of data processing, and (v) cost. We test these metrics in two bar-scale fluvial field environments: a large-river cobble bar and a sand-bedded river point bar at Neptune Beach State Park in western Oregon.

On the day of the survey, the conditions were very mixed, with periods of rain and wind interspersed with dry and calm periods, as well as some intermittent mist. This provided another means of comparing the approaches. Small 0.1 m targets were placed around the study areas for merging point clouds and to provide reference for later coordinate transforms, but the qualitative nature of this study did not require these targets other than for visual reference.

For each reach, the camera survey yielded about 60 photos taken in a sequential sweep pattern. The camera was held at eye height and hand triggered. For the Kinect survey, each reach required between 5 and 10 individual videos to be recorded with the device whilst slowly walking along the reach and panning the Kinect. Each recording was about 30 seconds in duration and about 1 Gb in size. To make longer recordings would have likely caused computational and memory constraint issues in the processing phase.

### Main findings

1 **Ease of use.** SfM-MVS provides a higher level of detail (Fig. B2.1i) and a simpler data collection scheme compared with the Kinect + laptop processor. The real-time feedback advantage of the Kinect ("live tracking") is not practical in the field, because we had to disable real-time tracking due to processor limitations. The need to break the video recording into 30-second clips becomes a logistical hassle and doubles the survey time as compared with SfM-MVS.

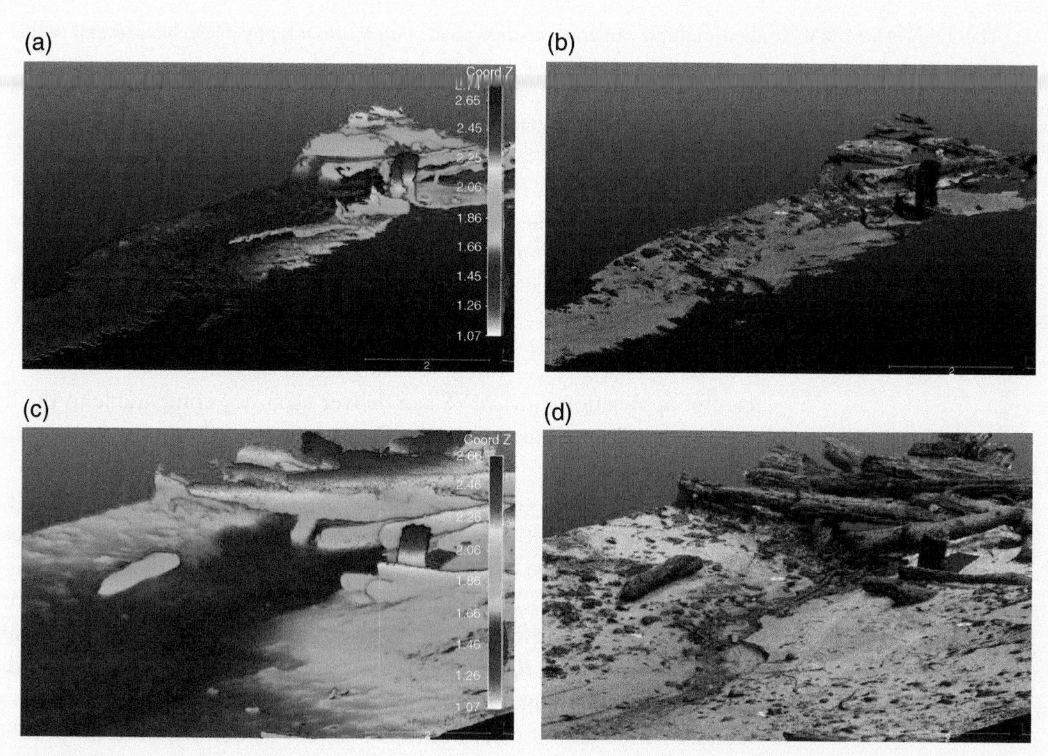


**Figure B2.1i** 3D points produced by Kinect and coloured by elevation (a); 3D points produced by Kinect and coloured by red–green–blue (RGB) attribute (b); 3D points produced by SfM-MVS and coloured by elevation (c); and 3D points produced by SfM-MVS and coloured by RGB attribute (d).

2. **Which amenable environments?** Neither SfM-MVS nor the Kinect was able to extract bathymetry from imaging at an eye level (Fig. B2.1i). Both approaches worked fairly well over sand, though the SfM-MVS approach seemed to work better in cobble and large wood jam situations.
3. **Robustness to changing conditions**. The complexity and exposed computing equipment of the Kinect system made it more difficult to use in rainy conditions. The changing lighting conditions did not appear to affect either approach.
4. **Ease of data processing**. In this scale range, SfM-MVS processing can be done as a single batch and is much simpler than the point cloud merging required by the Kinect approach. The resulting point clouds are denser from SfM-MVS. Low-angle photographs do sometimes cause camera position errors and need to be adjusted manually during the SfM-MVS processing phase.
5. **Cost?** The SfM-MVS approach used in this study costs us about GBP1120, though various options and costs could range from as little as GBP99 to a maximum of perhaps GBP2640 depending on camera choices and software licensing situations. The Kinect approach is less expensive than the SfM-MVS approach, with a basic cost of around GBP265 maximum.

### Key points for discussion

- At this time, the hardware and software performance of the Kinect has significant limitations that make it much more difficult to use in a river field situation in comparison with current SfM-MVS approaches.

- The ability of SfM-MVS to allow multiple cameras into individual scans means it is possible to have several people photographing a site at once, dramatically speeding up a survey.
- Both of these approaches, as well as new time-of-flight sensors and other advances, are likely to advance in the near future, and these conclusions should be considered provisional.

**Summary**

In river monitoring, river practitioners are asked to provide high-quality data in a small amount of time, often without complex technologies or large field crews. Both SfM-MVS and Kinect approaches provide for this need. From an operational standpoint, SfM is clearly a more useful river survey tool at the present time.

careful application, SfM-MVS can deliver accuracy comparable to the best achieved with any other topographic surveying method: direct or indirect (Table 2.3).

Box 2.1 contains a case study comparing SfM-MVS to Kinect, which is an active patterned-light measurement device, and in detail assessing each method with a range of metrics incorporating many practical aspects of surveying. If data processing costs are considered, substantial investment in personnel time, hardware and/or software is necessary to produce quality landform data using Kinect (Marcus & Fonstad 2008). Note when reading Box 2.1 that highlights "quality" must not be assumed to be the accuracy stated by the manufacturer but rather should be defined in terms of robust quantification of uncertainty of each point or derived (interpolated) grid cell in a surface, as depicted in Fig. 2.6.

#### 2.3.3.2 Challenges in Using SfM-MVS

SfM-MVS presents challenges for acquiring data over distances greater than 100 m. This challenge is due to the requirement for the camera position to move whilst maintaining a short distance to the landform of interest for the purpose of maintaining high-resolution images. However, mounting the camera on an airborne platform, which is the most common deployment method (Section 4.2.2), goes a long way to solving this problem and need not increase financial cost very much.

The 3D accuracy and 3D point density of SfM-MVS-derived 3D point data depends on factors outside of the control of the surveyor such as the texture and colour contrast of the landform of interest and ambient lighting conditions as discussed by Fonstad et al. (2013) and Gienko and Terry (2014), for example. This raises issues of repeatability and makes the SfM-MVS workflow challenging in conditions of poor illumination. Landforms that are highly reflective (glassy) cannot be reconstructed. Landforms that move within the time frame of acquiring images from different viewpoints (e.g. vegetation in the wind) cannot be surveyed using rigid SfM-MVS assumptions. Whilst georeferencing solutions using only non-specialist technology such as standard laser rangefinders are possible, accurate scaling and georeferencing require the (expert) use of a TS or dGPS.

Prevailing SfM-MVS workflows in the geosciences only produce data "back in the office." At the time of survey it cannot be known whether a point cloud will be successfully produced, or what attributes (extent, resolution, 3D point quality) that point cloud will have. Like any other point cloud data acquisition, SfM-MVS does not analyse the point cloud information automatically. In contrast, TS and dGPS permit the use of user-specified point codes, crucially at the time of point acquisition, to classify the type of point, to state, for example, that point represents a dune crest or a river bank crest.

SfM-MVS data at high resolution (sub-centimetre) or covering large spatial scales (>1000 $m^2$) has consequences in producing very large data sets, requiring large amounts of random access memory (RAM), and in producing long computational run times. However, these long run times are unlikely to be much more than the time that would otherwise have been spent acquiring the data by a different method. Moreover, SfM-MVS processing does not require continuous supervision by the operator. Advances in both CPU speeds and SfM-MVS-related algorithms will only improve in the future.

Efficient visualisation of SfM-MVS data is problematic. Industry-standard GISs are poor at handling large point cloud files. Software for the analysis of point cloud data is very much in its infancy. Consideration of visualisation of SfM-MVS-generated data on different platforms, such as mobile devices, has yet to be made. Chapter 4 provides further details.

## 2.4 Summary

This chapter first outlined the basic properties of different topographic survey platforms and digital sensors, indicating the survey design and methods that accompany each one. In doing so, this chapter has shown that each of the techniques outlined has different strengths and weaknesses and is better suited to different tasks. In particular, survey extent is one crucial consideration; it is not practical to consider applying ground-based techniques (TS, dGPS, TLS, SFM-MVS) at spatial extents that are trivial for aerial photogrammetry and ALS (>100's of $km^2$). Other equally important considerations are point density and 3D accuracy.

Furthermore, SfM-MVS is not a complete substitute for other digital surveying methods; accurate fluvial bathymetry over small areas is better obtained by TS or dGPS survey than SfM-MVS (Lejot et al. 2007; Bangen et al. 2014; Woodget et al. 2014), for example. However, in some circumstances, particularly where plot scale (~$10^1$ $m^2$) data is required (Smith & Vericat 2015) or where decimetre scale accuracy is acceptable over approximately 1 $km^2$ of bare ground (Javernick et al. 2014), SfM-MVS is an efficient and cost-effective survey method. In that respect, SfM-MVS contrasts with both ALS and TLS which are expensive survey solutions that produce large quantities of data which must then be decimated to produce useful terrain products.

In summary, the advantages of SfM-MVS in comparison to other digital survey methods are that:

- SfM-MVS is cheap.
- SfM-MVS is easy in terms of field data acquisition and operation of commercial or open-source software.
- SfM-MVS produces fully 3D data, as otherwise only possible with TLS.
- SfM-MVS derived point clouds and textured surfaces are inherently multi-dimensional ($x$, $y$, $z$, point orientation, colour, texture) and relatively easily transformed into orthophotographs and DEMs.
- SfM-MVS can be applied over a great range of spatial scales (examples to date are from $10^{-2}$ to $10^{6}\,m^2$, Smith & Vericat 2015), and that workflow remains virtually identical regardless of the spatial and temporal scales under consideration.
- With careful application, especially in the acquisition of GCPs, which require expert knowledge of TS or dGPS, SfM-MVS can deliver accuracy comparable to the best achieved with any other (direct or indirect) topographic surveying method.

The disadvantages of SfM-MVS in comparison to other digital survey methods are as follows:

- SfM-MVS presents challenges for acquiring and processing data over large spatial scales.
- The 3D accuracy of SfM-MVS-derived data depends on factors outside of the control of the surveyor, and this raises issues of repeatability.
- SfM-MVS only produces 3D data "back in the office" so at the time of survey it cannot be known whether a point cloud will be successfully produced or what attributes (extent, resolution, 3D point quality) that point cloud will have.
- SfM-MVS does not discriminate point locations or point types, nor does it analyse point cloud information automatically.
- Efficient visualisation of SfM-MVS data is problematic. Software for analysis of point cloud data is very much in its infancy.

Nonetheless, there are opportunities for SfM-MVS in the geosciences to mitigate some of these disadvantages by learning from developments in allied disciplines, and this is the subject matter of Chapter 7.

## References

Abellan, A., Oppikofer, T., Jaboyedoff, M., Rosser, N.J., Lim, M. & Lato, M.J. (2014) Terrestrial laser scanning of rock slope instabilities. *Earth Surface Processes and Landforms*, **39** (**1**), 80–97.

Alho, P., Kukko, A., Hyyppä, H., Kaartinen, H., Hyyppä, J. & Jaakkola, A. (2009) Application of boat-based laser scanning for river survey. *Earth Surface Processes and Landforms*, **34** (**13**), 1831–1838.

Baltsavias, E.P. (1999) Airborne laser scanning: basic relations and formulas. *ISPRS Journal of Photogrammetry and Remote Sensing*, **54** (**2**), 199–214.

Baltsavias, E.P., Favey, E., Bauder, A., Bosch, H. & Pateraki, M. (2001) Digital surface modelling by airborne laser scanning and digital photogrammetry for glacier monitoring. *The Photogrammetric Record*, **17** (**98**), 243–273.

Bangen, S.G., Wheaton, J.M., Bouwes, N., Bouwes, B. & Jordan, C. (2014) A methodological intercomparison of topographic survey techniques for characterizing wadeable streams and rivers. *Geomorphology*, **206**, 343–361.

Bemis, S.P., Micklethwaite, S., Turner, D. et al. (2014) Ground-based and UAV-based photogrammetry: A multi-scale, high-resolution mapping tool for structural geology and paleoseismology. *Journal of Structural Geology*, **69**, 163–178.

Bertoldi, W., Gurnell, A.M. & Drake, N.A. (2011) The topographic signature of vegetation development along a braided river: results of a combined analysis of airborne lidar, color air photographs, and ground measurements. *Water Resources Research*, **47** (**6**), W06525.

Bitelli, G., Dubbini, M. & Zanutta, A. (2004) Terrestrial laser scanning and digital photogrammetry techniques to monitor landslide bodies. *International Archives of Photogrammetry, Remote Sensing and Spatial Information Sciences*, **35** (**Part B5**), 246–251.

Brasington, J., Rumsby, B.T. & McVey, R.A. (2000) Monitoring and modelling morphological change in a braided gravel-bed river using high resolution GPS-based survey. *Earth Surface Processes and Landforms*, **25** (**9**), 973–990.

Brasington, J., Langham, J. & Rumsby, B. (2003) Methodological sensitivity of morphometric estimates of coarse fluvial sediment transport. *Geomorphology*, **53** (**3**), 299–316.

Brasington, J., Vericat, D. & Rychkov, I. (2012) Modeling river bed morphology, roughness, and surface sedimentology using high resolution terrestrial laser scanning. *Water Resources Research*, **48** (**11**), W11519.

Brown, R.A. & Pasternack, G.B. (2014) Hydrologic and topographic variability modulate channel change in mountain rivers. *Journal of Hydrology*, **510**, 551–564.

Carbonneau, P.E., Lane, S.N. & Bergeron, N.E. (2003) Cost-effective non-metric close-range digital photogrammetry and its application to a study of coarse gravel river beds. *International Journal of Remote Sensing*, **24** (**14**), 2837–2854.

Carbonneau, P.E., Lane, S.N. & Bergeron, N.E. (2004) Catchment-scale mapping of surface grain size in gravel bed rivers using airborne digital imagery. *Water Resources Research*, **40** (**7**), W07202.

Carrivick, J.L. & Twigg, D.R. (2005) Jökulhlaup-influenced topography and geomorphology at Kverkfjöll, Iceland. *Journal of Maps*, **1** (**1**), 7–17.

Carrivick, J.L., Russell, A.J., Rushmer, E.L. et al. (2009a) Geomorphological evidence towards a de-glacial control on volcanism. *Earth Surface Processes and Landforms*, **34** (**8**), 1164–1178.

Carrivick, J.L., Manville, V. & Cronin, S.J. (2009b) A fluid dynamics approach to modelling the 18th March 2007 lahar at Mt. Ruapehu, New Zealand. *Bulletin of Volcanology*, **71** (**2**), 153–169.

Carrivick, J.L., Manville, V., Graettinger, A. & Cronin, S.J. (2010) Coupled fluid dynamics-sediment transport modelling of a Crater Lake break-out lahar: Mt. Ruapehu, New Zealand. *Journal of Hydrology*, **388** (**3**), 399–413.

Carrivick, J.L., Smith, M.W., Quincey, D.J. & Carver, S.J. (2013a) Developments in budget remote sensing for the geosciences. *Geology Today*, **29** (**4**), 138–143.

Carrivick, J.L., Geilhausen, M., Warburton, J. et al. (2013b) Contemporary geomorphological activity throughout the proglacial area of an alpine catchment. *Geomorphology*, **188**, 83–95.

Carrivick, J.L., Turner, A.G., Russell, A.J., Ingeman-Nielsen, T. & Yde, J.C. (2013c) Outburst flood evolution at Russell Glacier, western Greenland: effects of a bedrock channel cascade with intermediary lakes. *Quaternary Science Reviews*, **67**, 39–58.

Carrivick, J.L., Smith, M.W. & Carrivick, D.M. (2015) Terrestrial laser scanning to deliver high-resolution topography of the upper Tarfala valley, arctic Sweden. *GFF*, 1–14, (ahead-of-print).

Casas, A., Benito, G., Thorndycraft, V.R. & Rico, M. (2006) The topographic data source of digital terrain models as a key element in the accuracy of hydraulic flood modelling. *Earth Surface Processes and Landforms*, **31** (**4**), 444–456.

Cavalli, M. & Tarolli, P. (2011) Application of LiDAR technology for rivers analysis. *Italian Journal of Engineering Geology and Environment*, **11**.

Chandler, J. (1999) Effective application of automated digital photogrammetry for geomorphological research. *Earth Surface Processes and Landforms*, **24** (**1**), 51–63.

Chandler, J., Ashmore, P., Paola, C., Gooch, M. & Varkaris, F. (2002) Monitoring river-channel change using terrestrial oblique digital imagery and automated digital photogrammetry. *Annals of the Association of American Geographers*, **92** (**4**), 631–644.

Charlton, M.E., Large, A.R. & Fuller, I.C. (2003) Application of airborne LiDAR in river environments: the River Coquet, Northumberland, UK. *Earth Surface Processes and Landforms*, **28** (**3**), 299–306.

Egli, L., Jonas, T., Grünewald, T., Schirmer, M. & Burlando, P. (2012) Dynamics of snow ablation in a small Alpine catchment observed by repeated terrestrial laser scans. *Hydrological Processes*, **26** (**10**), 1574–1585.

Fischer, L., Eisenbeiss, H., Kääb, A., Huggel, C. & Haeberli, W. (2011) Monitoring topographic changes in a periglacial high-mountain face using high-resolution DTMs, Monte Rosa East Face, Italian Alps. *Permafrost and Periglacial Processes*, **22** (**2**), 140–152.

Fonstad, M.A., Dietrich, J.T., Courville, B.C., Jensen, J.L. & Carbonneau, P.E. (2013) Topographic Structure from Motion: a new development in photogrammetric measurement. *Earth Surface Processes and Landforms*, **38** (**4**), 421–430.

Fuller, I.C., Large, A.R. & Milan, D.J. (2003) Quantifying channel development and sediment transfer following chute cutoff in a wandering gravel-bed river. *Geomorphology*, **54** (**3**), 307–323.

Gallay, M. (2013) Section 2.1.4: direct acquisition of data: airborne laser scanning. In: L.E. Clarke & J.M. Nield (eds), *Geomorphological Techniques (Online Edition)*. British Society for Geomorphology, London.

Gienko, G.A. & Terry, J.P. (2014) Three-dimensional modeling of coastal boulders using multi-view image measurements. *Earth Surface Processes and Landforms*, **39** (7), 853–864.

Hakala, T., Suomalainen, J., Kaasalainen, S. & Chen, Y. (2012) Full waveform hyperspectral LiDAR for terrestrial laser scanning. *Optics Express*, **20** (7), 7119–7127.

Heritage, G.L. & Milan, D.J. (2009) Terrestrial laser scanning of grain roughness in a gravel-bed river. *Geomorphology*, **113** (**1**), 4–11.

Hilldale, R.C. & Raff, D. (2008) Assessing the ability of airborne LiDAR to map river bathymetry. *Earth Surface Processes and Landforms*, **33** (**5**), 773–783.

Hodge, R., Brasington, J. & Richards, K. (2009) In situ characterization of grain-scale fluvial morphology using Terrestrial Laser Scanning. *Earth Surface Processes and Landforms*, **34** (7), 954–968.

Hodgson, M.E. & Bresnahan, P. (2004) Accuracy of airborne LiDAR-derived elevation: empirical assessment and error budget. *Photogrammetric Engineering & Remote Sensing*, **70** (**3**), 331–339.

Hollaus, M., Mücke, W., Roncat, A., Pfeifer, N. & Briese, C. (2014) Full-waveform airborne laser scanning systems and their possibilities in forest applications. In: M. Maltamo, E. Naesset & J. Vauhkonen (eds), *Forestry Applications of Airborne Laser Scanning*, pp. 43–61. Springer, Dordrecht.

Hugenholtz, C.H., Whitehead, K., Brown, O.W. et al. (2013) Geomorphological mapping with a small unmanned aircraft system (sUAS): feature detection and accuracy assessment of a photogrammetrically-derived digital terrain model. *Geomorphology*, **194**, 16–24.

Irish, J.L. & Lillycrop, W.J. (1999) Scanning laser mapping of the coastal zone: the SHOALS system. *ISPRS Journal of Photogrammetry and Remote Sensing*, **54** (**2**), 123–129.

Javernick, L., Brasington, J. & Caruso, B. (2014) Modeling the topography of shallow braided rivers using Structure-from-Motion photogrammetry. *Geomorphology*, **213**, 166–182.

Joerg, P.C., Morsdorf, F. & Zemp, M. (2012) Uncertainty assessment of multi-temporal airborne laser scanning data: a case study on an Alpine glacier. *Remote Sensing of Environment*, **127**, 118–129.

Jones, A.F., Brewer, P.A., Johnstone, E. & Macklin, M.G. (2007) High-resolution interpretative geomorphological mapping of river valley environments using airborne LiDAR data. *Earth Surface Processes and Landforms*, **32** (**10**), 1574–1592.

Kankare, V., Holopainen, M., Vastaranta, M. et al. (2013) Individual tree biomass estimation using terrestrial laser scanning. *ISPRS Journal of Photogrammetry and Remote Sensing*, **75**, 64–75.

Keim, R.F., Skaugset, A.E. & Bateman, D.S. (1999) Digital terrain modeling of small stream channels with a total-station theodolite. *Advances in Water Resources*, **23** (**1**), 41–48.

Knoll, C. & Kerschner, H. (2010) A glacier inventory for South Tyrol, Italy, based on airborne laser-scanner data. *Annals of Glaciology*, **50** (**53**), 46–52.

Kreylos, O., Oskin, M., Cowgill, E., Gold, P., Elliott, A. & Kellogg, L. (2013) Point-based computing on scanned terrain with LidarViewer. *Geosphere*, **9** (**3**), 546–556.

Lane, S.N., Richards, K.S. & Chandler, J.H. (1994) Developments in monitoring and modelling small-scale river bed topography. *Earth Surface Processes and Landforms*, **19** (**4**), 349–368.

Lane, S.N., James, T.D. & Crowell, M.D. (2000) Application of digital photogrammetry to complex topography for geomorphological research. *The Photogrammetric Record*, **16** (**95**), 793–821.

Lane, S.N., Widdison, P.E., Thomas, R.E. et al. (2010) Quantification of braided river channel change using archival digital image analysis. *Earth Surface Processes and Landforms*, **35** (**8**), 971–985.

Lejot, J., Delacourt, C., Piégay, H., Fournier, T., Trémélo, M.L. & Allemand, P. (2007) Very high spatial resolution imagery for channel bathymetry and topography from an unmanned mapping controlled platform. *Earth Surface Processes and Landforms*, **32** (**11**), 1705–1725.

Liang, X., Litkey, P., Hyyppa, J., Kaartinen, H., Vastaranta, M. & Holopainen, M. (2012) Automatic stem mapping using single-scan terrestrial laser scanning. *IEEE Transactions on Geoscience and Remote Sensing*, **50** (**2**), 661–670.

Lim, M., Petley, D.N., Rosser, N.J., Allison, R.J., Long, A.J. & Pybus, D. (2005) Combined digital photogrammetry and time-of-flight laser scanning for monitoring cliff evolution. *The Photogrammetric Record*, **20** (**110**), 109–129.

Lin, Z., Kaneda, H., Mukoyama, S., Asada, N. & Chiba, T. (2013) Detection of subtle tectonic–geomorphic features in densely forested mountains by very high-resolution airborne LiDAR survey. *Geomorphology*, **182**, 104–115.

Lindberg, E., Olofsson, K., Holmgren, J. & Olsson, H. (2012) Estimation of 3D vegetation structure from waveform and discrete return airborne laser scanning data. *Remote Sensing of Environment*, **118**, 151–161.

Mallet, C. & Bretar, F. (2009) Full-waveform topographic lidar: state-of-the-art. *ISPRS Journal of Photogrammetry and Remote Sensing*, **64** (**1**), 1–16.

Marcus, W.A. & Fonstad, M.A. (2008) Optical remote mapping of rivers at sub-meter resolutions and watershed extents. *Earth Surface Processes and Landforms*, **33** (**1**), 4–24.

McCoy, S.W., Kean, J.W., Coe, J.A., Staley, D.M., Wasklewicz, T.A. & Tucker, G.E. (2010) Evolution of a natural debris flow: in situ measurements of flow dynamics, video imagery, and terrestrial laser scanning. *Geology*, **38** (**8**), 735–738.

Milan, D.J., Heritage, G.L. & Hetherington, D. (2007) Application of a 3D laser scanner in the assessment of erosion and deposition volumes and channel change in a proglacial river. *Earth Surface Processes and Landforms*, **32** (**11**), 1657–1674.

Mora, P., Baldi, P., Casula, G. et al. (2003) Global positioning systems and digital photogrammetry for the monitoring of mass movements: application to the Ca'di Malta landslide (northern Apennines, Italy). *Engineering Geology*, **68** (**1**), 103–121.

Nelson, C.E., Jerram, D.A., Hobbs, R.W., Terrington, R. & Kessler, H. (2011) Reconstructing flood basalt lava flows in three dimensions using terrestrial laser scanning. *Geosphere*, **7** (**1**), 87–96.

Nield, J.M., Wiggs, G.F. & Squirrell, R.S. (2011) Aeolian sand strip mobility and protodune development on a drying beach: examining surface moisture and surface roughness patterns measured by terrestrial laser scanning. *Earth Surface Processes and Landforms*, **36** (**4**), 513–522.

Noh, M.J. & Howat, I.M. (2015) Automated stereo-photogrammetric DEM generation at high latitudes: surface extraction with TIN-based Search-space Minimization (SETSM) validation and demonstration over glaciated regions. *GIScience & Remote Sensing*, **52** (**2**), 198–217.

Notebaert, B., Verstraeten, G., Govers, G. & Poesen, J. (2009) Qualitative and quantitative applications of LiDAR imagery in fluvial geomorphology. *Earth Surface Processes and Landforms*, **34** (**2**), 217–231.

Razak, K.A., Santangelo, M., Van Westen, C.J., Straatsma, M.W. & de Jong, S.M. (2013) Generating an optimal DTM from airborne laser scanning data for landslide mapping in a tropical forest environment. *Geomorphology*, **190**, 112–125.

Ribeiro, R.D.R., Ramirez, E., Simoes, J.C. & Machaca, A. (2013) 46 years of environmental records from the Nevado Illimani glacier group, Bolivia, using digital photogrammetry. *Annals of Glaciology*, **54** (**63**), 272–278.

Rosser, N.J., Petley, D.N., Lim, M., Dunning, S.A. & Allison, R.J. (2005) Terrestrial laser scanning for monitoring the process of hard rock coastal cliff erosion. *Quarterly Journal of Engineering Geology and Hydrogeology*, **38** (**4**), 363–375.

Rychkov, I., Brasington, J. & Vericat, D. (2012) Computational and methodological aspects of terrestrial surface analysis based on point clouds. *Computers and Geosciences*, **42**, 64–70.

Sallenger, A.H., Jr, Krabill, W.B., Swift, R.N. et al. (2003) Evaluation of airborne topographic lidar for quantifying beach changes. *Journal of Coastal Research*, **19**, 125–133.

Slatton, K.C., Carter, W.E., Shrestha, R.L. & Dietrich, W. (2007) Airborne laser swath mapping: achieving the resolution and accuracy required for geosurficial research. *Geophysical Research Letters*, **34** (**23**), L23S10.

Smith, M.W. (2015) Section 2.1.5 direct acquisition of elevation data: terrestrial laser scanning. In: L.E. Clarke & J.M. Nield (eds), *Geomorphological Techniques (Online Edition)*. British Society for Geomorphology, London.

Smith, M.W. & Vericat, D. (2015) From experimental plots to experimental landscapes: topography, erosion and deposition in sub-humid badlands from Structure-from-Motion photogrammetry. *Earth Surface Processes and Landforms*, **40**, 1656–1671.

Smith, M.W., Cox, N.J. & Bracken, L.J. (2011) Terrestrial laser scanning soil surfaces: a field methodology to examine soil surface roughness and overland flow hydraulics. *Hydrological Processes*, **25** (**6**), 842–860.

Smith, M.W., Vericat, D. & Gibbins, C. (2012) Through-water terrestrial laser scanning of gravel beds at the patch scale. *Earth Surface Processes and Landforms*, **37** (**4**), 411–421.

Srinivasan, S., Popescu, S.C., Eriksson, M., Sheridan, R.D. & Ku, N.W. (2014) Multi-temporal terrestrial laser scanning for modeling tree biomass change. *Forest Ecology and Management*, **318**, 304–317.

Staines, K.E., Carrivick, J.L., Tweed, F.S. et al. (2014) A multi-dimensional analysis of proglacial landscape change at Sólheimajökull, southern Iceland. *Earth Surface Processes and Landforms*, **40**, 809–822.

Sturzenegger, M. & Stead, D. (2009) Close-range terrestrial digital photogrammetry and terrestrial laser scanning for discontinuity characterization on rock cuts. *Engineering Geology*, **106** (**3**), 163–182.

Tsai, Z.X., You, G.J.Y., Lee, H.Y. & Chiu, Y.J. (2012) Use of a total station to monitor post-failure sediment yields in landslide sites of the Shihmen reservoir watershed, Taiwan. *Geomorphology*, **139**, 438–451.

Vallé, B.L. & Pasternack, G.B. (2006) Field mapping and digital elevation modelling of submerged and unsubmerged hydraulic jump regions in a bedrock step–pool channel. *Earth Surface Processes and Landforms*, **31** (**6**), 646–664.

Wagner, W., Hollaus, M., Briese, C. & Ducic, V. (2008) 3D vegetation mapping using small-footprint full-waveform airborne laser scanners. *International Journal of Remote Sensing*, **29** (**5**), 1433–1452.

Westoby, M.J., Brasington, J., Glasser, N.F., Hambrey, M.J. & Reynolds, J.M. (2012) "Structure-from-Motion" photogrammetry: a low-cost, effective tool for geoscience applications. *Geomorphology*, **179**, 300–314.

Wheaton, J.M., Brasington, J., Darby, S.E. & Sear, D.A. (2010) Accounting for uncertainty in DEMs from repeat topographic surveys: improved sediment budgets. *Earth Surface Processes and Landforms*, **35** (**2**), 136–156.

Williams, R.D., Brasington, J., Vericat, D. & Hicks, D.M. (2014) Hyperscale terrain modelling of braided rivers: fusing mobile terrestrial laser scanning and optical bathymetric mapping. *Earth Surface Processes and Landforms*, **39** (**2**), 167–183.

Wood, J. (2009) Geomorphometry in LandSerf. In: T. Hengl & H.I. Reuter (eds), *Geomorphometry—Concepts, Software, Applications, Developments in Soil Science*, **33**, pp. 333–349. Elsevier, Amsterdam.

Yang, X., Strahler, A.H., Schaaf, C.B. et al. (2013) Three-dimensional forest reconstruction and structural parameter retrievals using a terrestrial full-waveform lidar instrument (Echidna®). *Remote Sensing of Environment*, **135**, 36–51.

Young, E.J. (2013) Section 2.1.3: dGPS. In: S.J. Cook, L.E. Clarke & J.M. Nield (eds), *Geomorphological Techniques (Online Edition)*. British Society for Geomorphology, London.

## Further Reading/Resources

The British Society for Geomorphology is developing a free, peer-reviewed online publication "Geomorphological Techniques" (hyperlink: http://www.geomorphology.org.uk/onsite_publications), which includes dGPS and TLS topographic surveying methods. It is an evolving resource and has more information per method than could be covered in this chapter.

# 3 Background to Structure from Motion

Abstract
Structure from Motion (SfM) as applied in the geosciences is not so much a single technique as a workflow employing multiple algorithms developed from three-dimensional (3D) computer vision, traditional photogrammetry, and more conventional survey techniques. Only one of these steps is technically "SfM" as defined in the computer vision literature. In full, the workflow is commonly known as SfM-MVS, to account for the Multi-View Stereo (MVS) algorithms used in the final stages. This chapter outlines each step in the workflow, namely (i) detecting image features or keypoints, (ii) identifying correspondences between these keypoints on different images, (iii) filtering these links to remove geometrically inconsistent keypoint correspondences, (iv) "SfM" or simultaneously estimating 3D scene geometry, camera pose, and internal camera parameters through a bundle adjustment, (v) scaling and georeferencing the resultant scene geometry, (vi) optimising the parameters identified in the bundle adjustment using known ground control points (GCPs), (vi) clustering image sets for efficient processing, and (vii) applying MVS algorithms. In this chapter, algorithms and assumptions underlying each of the aforementioned steps are described accessibly in order to allow the typical geoscience user to be more informed on the processes and assumptions taking place within their chosen software package.

Keywords
computer vision; keypoint; SIFT; bundle adjustment; Multi-View stereo; photogrammetry

## 3.1 Introduction

This chapter provides deeper background on the operation of Structure from Motion–Multi-View Stereo (SfM-MVS). The main concepts underlying the method are presented, but detailed mathematical formulae are avoided

*Structure from Motion in the Geosciences*, First Edition. Jonathan L. Carrivick, Mark W. Smith, and Duncan J. Quincey.

Companion Website: www.wiley.com/go/carrivick/structuremotiongeosciences

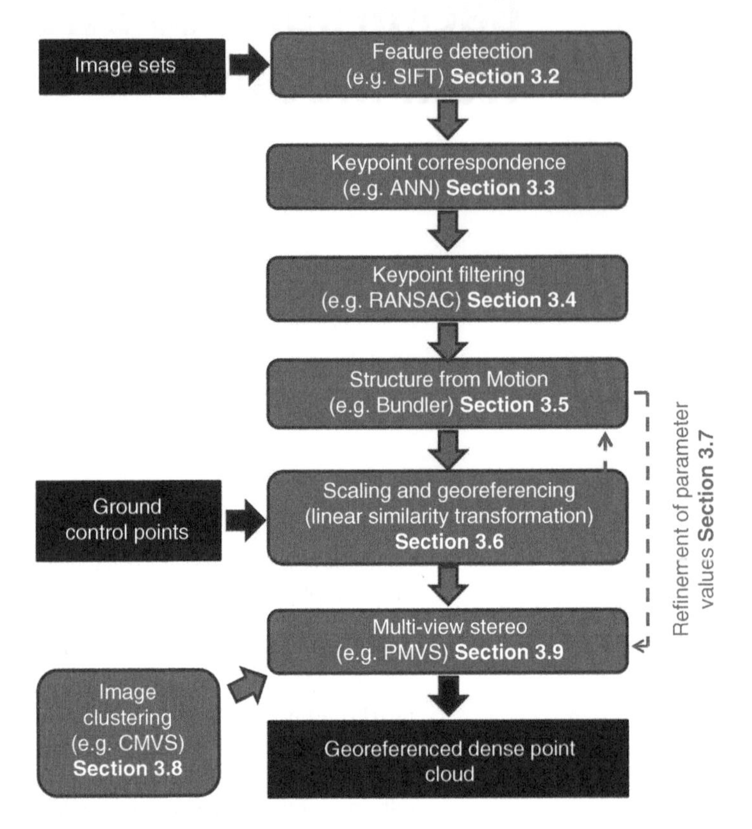


**Figure 3.1** Typical workflow in the production of georeferenced dense point clouds from image sets and ground control points. The workflow provides the structure for this chapter. Inputs and outputs are shown in red.

intentionally; rather, the aim is to present a typical geoscience audience with an understanding of different processes involved in the reconstruction of three-dimensional (3D) geometry from a sequence of standard uncalibrated imagery (summarised in Fig. 3.1). Readers interested in the associated mathematical equations are directed to relevant sources for this information (e.g. Triggs et al. 2000; Hartley & Zisserman 2003; Lowe 2004; Snavely 2008; Szeliski 2011).

Traditional photogrammetry is long established in the geosciences. Yet, as described in Section 2.3.1, several elements of this technique are rather restrictive. For example, there is a requirement for near-parallel stereopairs of images with approximately 60 % overlap, accurate measurement of the 3D location and pose of the camera for each image is necessary, either directly or resectioned from control points for the estimation of the camera pose. Camera calibration is essential because there is a lack of redundancy as both images must contribute to the final DEM. These elements result in a time-consuming process with a steep learning curve.

More flexible photogrammetric methods have since emerged with soft-copy photogrammetry now used widely by the geoscience community, enabling the derivation of high-quality topographic data products over a wide range of spatial scales.

Given these relatively recent developments in traditional photogrammetry, it is tempting to assume that SfM-MVS simply represents the latest incremental development of photogrammetry. Yet, this is not strictly the case. SfM-MVS merges photogrammetric principles with developments of an entirely different origin, coming from advances in 3D computer vision algorithms since the 1980s. With the proliferation of digital photography in the late 1990s and increased availability of imagery, applications of SfM-MVS algorithms have become more readily apparent.

The following description of the SfM-MVS approach describes the typical processes and workflow required to reconstruct the 3D scene geometry from a set of images where the extrinsic and intrinsic calibration parameters are unknown (see Fig. 3.1 for an overview). There are actually many alternative SfM-based approaches to this problem; each particular software implementation of SfM-MVS will be slightly different. With this variability in mind and considering that many commercial SfM-MVS software packages do not detail the specific procedure applied, this chapter does not attempt to detail every available approach. In general, the details that are given in this chapter are based on the approach detailed in Snavely et al. (2008) because that is open source and covers many elements common with other approaches. We do, however, also include information on alternative approaches or additional processing steps not described by Snavely et al. (2008) in order to provide a comprehensive and up-to-date overview of the technique as applied in the geosciences. For note, details of specific software packages are provided in Chapter 4.

## 3.2 Feature Detection

In the past few decades, several slightly different advances in feature detection were made in parallel in both fields of digital image processing and computer vision. The latter also aimed to then recover the 3D structure of the images. The fundamental question driving the development of both approaches was how to best extract descriptions of local points in a way that allowed the correct identification of correspondences between those points (often from a large dataset of such points), but that was insensitive to changes in orientation, scale, illumination, or 3D position (Fig. 3.2).

The first step of image matching involves the identification of common points on a number of different photographs. It is these **keypoints** that allow the different images to be matched and the scene geometry reconstructed. A number of techniques to identify keypoints have been developed based either on matching image statistics (Lucas & Kanade 1981), identifying "corner-like" features (Moravec 1983) from large gradients in all directions, or later by using eigenvalues of smoothed outer products of gradients (Förstner 1986; Harris & Stephens 1988). Initial applications of keypoint matching were for stereo and short-range motion tracking. However, as discussed by Snavely (2008), these techniques were limited to the identification of keypoints at a single scale and were only applicable to scenes with images taken from a similar viewpoint. The techniques listed earlier do

**Figure 3.2** Example of the challenges facing feature matching algorithms. Keypoints need to be matched on images with variable 3D position, scale, orientation, and illumination as demonstrated in these two images of an abandoned aqueduct.

work where features can be tracked from one frame to the next, but the challenge of identifying features that can be tracked between images taken from widely different views (so-called wide baseline matching, e.g. Baumberg 2000; Matas et al. 2004) requires application of alternative techniques.

**Feature points** (i.e. sets of pixels) that are invariant to changes in scale and orientation and that are affine invariant (or *covariant* regions, changing covariantly with the transformation; Mikolajczyk et al. 2005) are required for wide baseline matching. The shape of the region of interest has to be able to adapt to cope with geometric distortions in the target feature owing to a change in perspective between two images. This accommodation of geometric distortion is demonstrated in Fig. 3.3 (from Mikolajczyk et al. 2005) where elliptical regions are detected independently in each viewpoint but correspond to the same surface region. Geometric and photometric deformations can be normalised to obtain viewpoint- and illumination-invariant descriptions of the intensity pattern in the region (Mikolajczyk et al. 2005).

Whilst many different region detectors and feature types are available (the performance of several is compared by Mikolajczyk et al. (2005)), the scale-invariant feature transform (SIFT) object-recognition system (Lowe 1999, 2001, 2004) is used most widely. SIFT allows the relative position of the feature to shift dramatically with only small changes in the descriptor. Furthermore, SIFT is robust against changes in 3D viewpoint for non-planar surfaces (Lowe 2004). However, SIFT is not fully affine invariant, which would be more useful for matching planar surfaces under large view changes (Fig. 3.3). Lowe (2004) suggested that a combination of SIFT with other feature types would provide further image matches under different circumstances and that would likely be implemented in future systems. Indeed, the

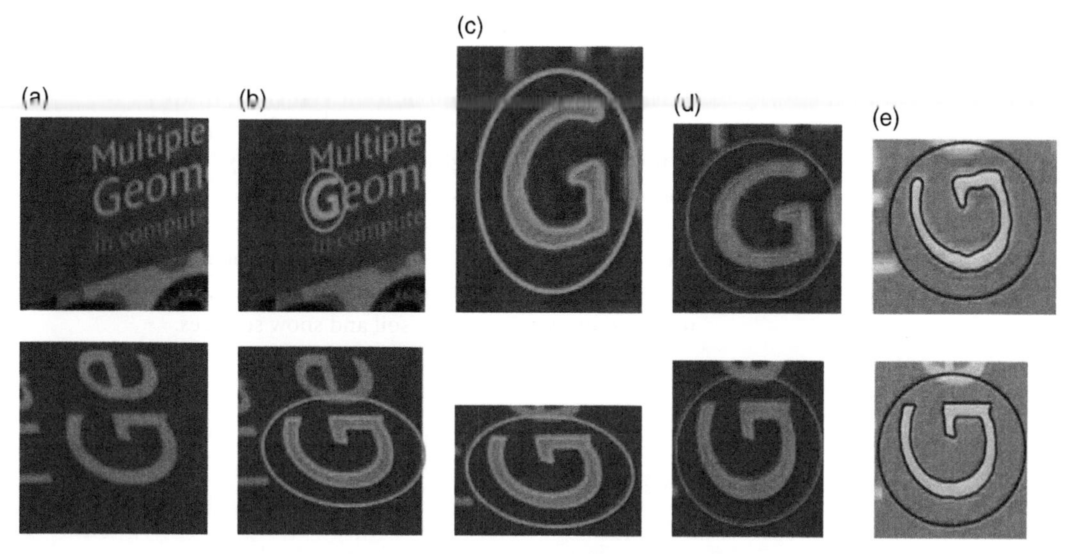


**Figure 3.3** Affine covariant regions as a solution to variable viewpoint and illumination. Two separate viewpoints are demonstrated (one per row). Original images (a) from which affine covariant regions are detected (b) shown in close-up (c). Regions are geometrically normalised to circles (d) and are the same up to rotation. Geometric and photometric normalisation (e) leaves only slight residual differences in rotation due to estimation error. The two features in the images in (e) are thus likely to be matched following these transformations. Source: From Mikolajczyk et al. (2005, p. 45).

conclusion of Mikolajczyk et al. (2005) was that the use of several complementary detectors should be used to extract regions with different properties.

Owing to the popularity of the SIFT object-recognition system, it is described in detail here, but for further information interested readers are directed to Lowe (2004) from which this summary is taken. The SIFT code is available from http://www.cs.ubc.ca/~lowe/keypoints/. In short, SIFT follows four major stages:

1. **Detection of spatial extrema.** This first step of SIFT involves an efficient identification of locations and scales that can be assigned repeatedly to the same object from differing viewpoints. A space-scale approach is used which detects locations that are invariant to scale changes by searching for stable features across a continuous function of scale. A monochrome intensity image is convolved with a Gaussian function incrementally at different scales and the difference between consecutive Gaussian images subtracted. Local extrema are then detected by comparing each sample point with its eight neighbours in the current image and nine neighbours in the scales above and below. Lowe (2004) analyzed the preferred sampling frequency in both scale and space, suggesting that most of the stable and useful features could be detected with coarse sampling.
2. **Keypoint localisation.** SIFT then performs a detailed fit of a 3D quadratic function for each candidate keypoint to nearby data for

location, scale, and ratio of principal curvatures. Large numbers of keypoints are typically identified (Fig. 3.4b). Rejected points may have low contrast (removed from Fig. 3.4c) or are poorly localised along an edge (reflected in a high ratio of principal curvatures; removed from Fig. 3.4d). The density of keypoints identified in an image depends on the texture, sharpness, and resolution of the image. Complex scenes will work best, whilst relatively featureless surfaces such as snow and sand are likely to prove the most challenging. For an illustration of keypoint localisation, Fig. 3.5 compares keypoint matches for image pairs on soil and snow surfaces.

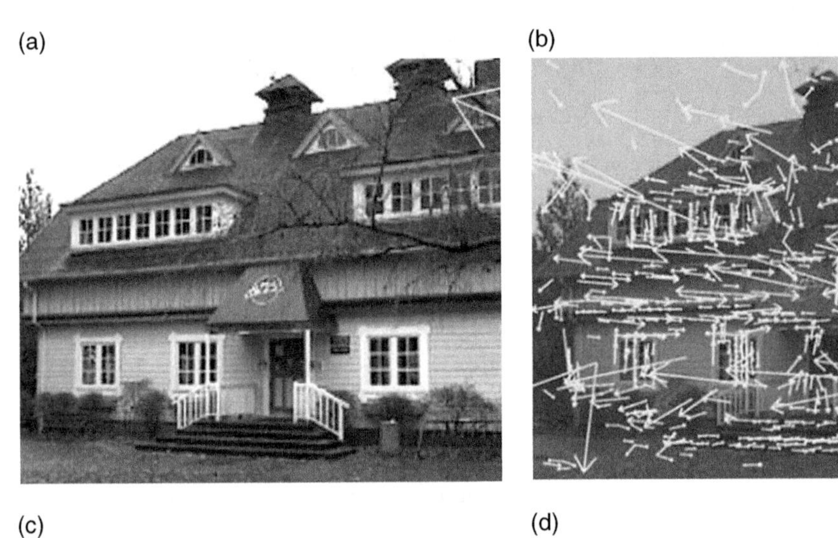

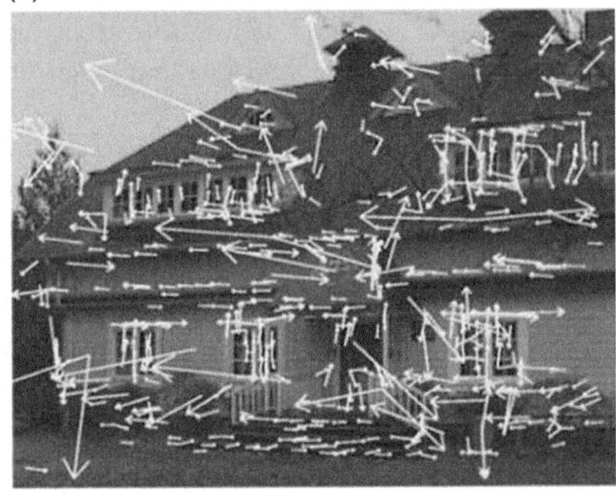


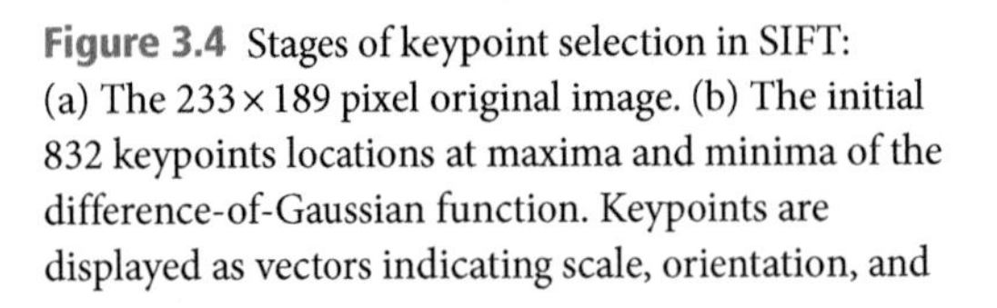

**Figure 3.4** Stages of keypoint selection in SIFT: (a) The 233 × 189 pixel original image. (b) The initial 832 keypoints locations at maxima and minima of the difference-of-Gaussian function. Keypoints are displayed as vectors indicating scale, orientation, and location. (c) After applying a threshold on minimum contrast, 729 keypoints remain. (d) The final 536 keypoints that remain following an additional threshold on the ratio of principal curvatures. Source: From Lowe (2004, p. 98).

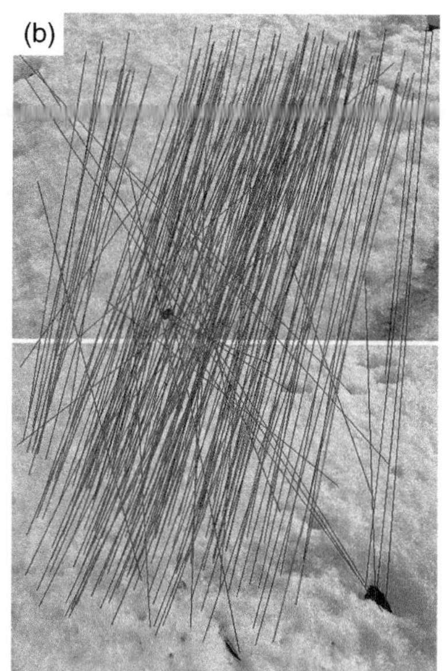


**Figure 3.5** Tracks between two images for plot-scale SfM-MVS surveys of (a) a stony soil surface and (b) a snow surface. Invalid matches are shown in red, whilst valid matches are given in blue. A stony soil surface has many more matches than a relatively texture-free ice surface. The number of matches is strongly influenced by relative image position. An attempt has been made to standardise this here, but the greater density of keypoint matches in (a) is broadly representative of the difference between the two surfaces. Matches were found and filtered using Agisoft PhotoScan.

3 **Orientation assignment**. A consistent orientation for each keypoint is assigned through analysis of dominant directions of local intensity gradients using the Gaussian-smoothed image closest to the scale of the keypoint. Where a second peak in the orientation histogram is identified (within 80% of the highest peak), a second keypoint is created at that location and scale but with a different orientation.
4 **Keypoint descriptor**. Next, a descriptor for each keypoint is required that is sufficiently distinctive yet is as invariant as possible to changes in 3D viewpoint or illumination. The approach of SIFT to gain this descriptor is that it considers sensitivity to intensity gradients but not the location of those gradients. Gradient magnitudes and orientations are sampled around each keypoint, rotated relative to the keypoint orientation (Fig. 3.6). A Gaussian weighting function is applied to these gradients to avoid large gradients far from the centre of the descriptor determining the specific descriptor. Gradients are accumulated into orientation histograms over $4 \times 4$ sample regions (thereby permitting local positional shifts). The descriptor is thus a $4 \times 4$ array of histograms with eight orientation bins, each resulting in a 128-element feature vector for each keypoint. To avoid illumination effects, the vector is normalised to unit length, thereby correcting

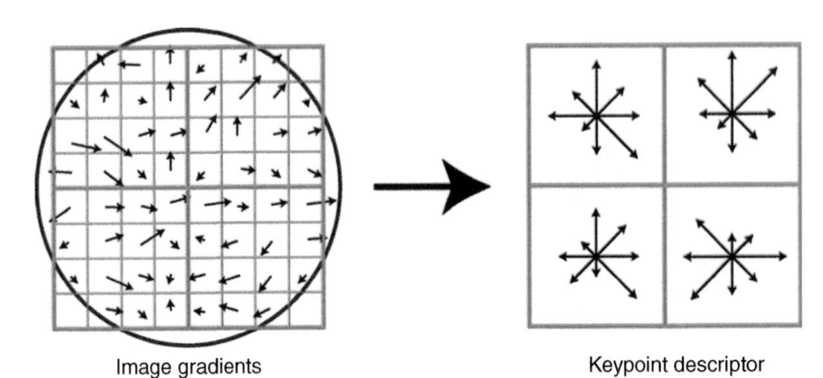


**Figure 3.6** Example of a keypoint descriptor. First the gradient magnitude and orientation at each image sample point in a region around the keypoint location is calculated (left). These are weighted by a Gaussian window, indicated by the overlaid circle. These samples are then accumulated into orientation histograms summarising the contents over $4 \times 4$ sub-regions (right) with the length of each arrow corresponding to the sum of the gradient magnitudes near that direction within the region. This figure demonstrates a $2 \times 2$ descriptor array computed from an $8 \times 8$ set of samples. Source: From Lowe (2004, p. 101).

for contrast changes. In addition, values in the unit feature vector are thresholded to avoid large gradient magnitudes effecting the matching and thereby placing greater focus on vector orientations (that are less sensitive to non-linear illumination changes). The resulting descriptor has been shown to discriminate individual keypoints from large databases (of tens of thousands).

A comparative study of other view-invariant local image descriptors by Mikolajczyk and Schmid (2005) found that a variant of SIFT (the gradient location-orientation histogram, GLOH) outperformed SIFT but by only a small margin. GLOH differs from SIFT in that it uses log-polar bins instead of square bins to compute the orientation histograms. For note, other subsequent variations include SURF (Bay et al. 2008), ASIFT (Morel & Yu 2009), BRIEF (Calonder et al. 2010) and LDAHash (Strecha et al. 2012).

## 3.3 Keypoint Correspondence

Once keypoints have been located in each image, correspondences between keypoints in different images need to be determined. Yet, there is no guarantee that any given keypoint will have a partner in another image. Therefore, methods for discarding points with no good match are required. Working with the 128-dimension keypoint data of the SIFT algorithm, Lowe (2004) used the ratio of the Euclidean distance of the nearest neighbour with that of the second nearest, specifying a minimum value of 0.8 (though Snavely et al. (2008) lower the threshold to 0.6). This "distance ratio" criterion was observed to eliminate 90% of false matches whilst discarding only less than

5% of correct matches. It was also found to perform better than a global distance threshold as the distance ratio specifies that correct matches must be substantially "more correct" than the other options, increasing the likelihood of a reliable match. Conversely, spurious matches are unlikely to be much better than the second closest incorrect match given the high dimensionality of the feature space (Lowe 2004).

The complexity of keypoint descriptors and the typically large number of keypoints mean that performing an exhaustive brute-force Euclidean nearest neighbour search in such a high-dimensional space is both difficult and computationally expensive (Arya et al. 1998). An efficient solution to this problem has been ***k*-dimensional trees** (or *k-d* trees), which are a type of binary tree often used to space-partition multi-dimensional data for nearest neighbour calculation (Bentley 1975; Friedman et al. 1977). At each level, *k-d* trees partition data points into bins using a different dimension, often splitting the data using the median value as a splitting point. The resulting nearest neighbour search works recursively, and the advantage of the data structure is that it quickly eliminates large regions of the search space.

Nonetheless, owing to the "curse of dimensionality," problems arise with the high-dimensional spaces of complex keypoint descriptors (e.g. the SIFT keypoint descriptor has 128 dimensions). In general, where the dimensionality is $k$, the efficiency gains of *k-d* trees is no better than an exhaustive search unless the number of data points $N >> 2^k$. In practice, where $k > 8$, the computation of nearest neighbours requires modification to allow for "approximate matching" in which non-optimal neighbours are sometimes identified in return for an order of magnitude search time improvements (Muja & Lowe 2009). Arya et al. (1998) achieve this by modifying the *k-d* tree algorithm by permitting the identified nearest neighbour of a point to be within a relative error. They also implement a priority search algorithm where bins of the tree are searched in order, starting with those where the feature space is closest to the query point location. Similarly, the "best-bin-first" (BBF) approach of Beis and Lowe (1997) implements a priority search order but sets a limit on the number of bins visited in the tree. For note, alternative approaches to the approximate matching problem are described in Muja and Lowe (2009). Implementing high-dimensional nearest neighbour searches on a graphics processing unit (GPU) has also been shown to decrease the required search times (Bustos et al. 2006) running six times faster.

Lowe (2004) notes that cutting off the approximate nearest neighbour (ANN) search after only checking the first 200 nearest-neighbour candidates provides a large time saving (two orders of magnitude where keypoints number >100,000) whilst only losing less than 5% of correct matches. Therefore, when coupled with the distance ratio criterion described earlier, the BBF algorithm need not provide exact solutions of the most difficult cases of keypoints with many close neighbours as the correspondence would be rejected by the distance ratio criterion in any case.

## 3.4 Identifying Geometrically Consistent Matches

To be confident that only correct correspondences remain, a further step is applied to filter out any remaining erroneous matches. With multiple keypoints identified in a pair of images of the same scene (e.g. Fig. 3.5), the **fundamental matrix** for the image pair is calculated (similar to the essential matrix but for uncalibrated cameras). By specifying the relationship between the two images, the fundamental matrix (or F-matrix) constrains the locations of correctly identified keypoints in both images and can be calculated using the eight-point algorithm (Longuet-Higgins 1981). This algorithm uses eight-point matches (or more) on two uncalibrated views, and a set of linear equations to reconstruct a scene up to a projective transformation where all points lying on a single line will remain aligned in this way (i.e. "collinearity" is preserved). The eight-point algorithm is a simple and rapid approach to compute the F-matrix, but it is sensitive to noise in the specified location of keypoints. Simple normalisation of the points in the image prior to solving the set of linear equations has been shown by Hartley (1997) to yield a large improvement in the use of the eight-point algorithm. Alternatively, more complex approaches to identify the F-matrix have been developed; see Zhang (1996) for details.

Candidate F-matrices are calculated over several iterations using either the least median of squares or, more commonly, the random sample consensus (RANSAC) method (Fischler & Bolles 1981). RANSAC is the more popular method because it is fast, accurate, and robust (Sunglok et al. 2009). The RANSAC method assumes that all keypoints can be divided into two sets: outliers and inliers. A perfect model fit would ignore all outliers and would be computed exclusively from inliers. An initial random sample of the keypoints is taken from which the F-matrix is calculated using the smallest possible subset of the data, in this case seven points. The error of each keypoint with respect to the estimation is then computed, and the number of inlier points counted. To define inliers, some threshold must be specified. Snavely et al. (2008) set this threshold to be 0.6% of the maximum image dimension, typically six pixels for an $1024 \times 768$ image. The sampling process is repeated on different subsets iteratively for a sufficient number of times to ensure that there is a 95% chance that one of the subsets contains only "inliers." RANSAC then returns the F-matrix with the largest number of inliers. The final model is then computed using only the inlier set. The robustness of the RANSAC method to outliers in comparison to least squares methods is demonstrated with a simple example in Fig. 3.7.

Snavely et al. (2008) further refine this F-matrix by running the iterative Levenberg–Marquardt (LM) algorithm on the inlier set (Levenberg 1944; Marquardt 1963). The LM algorithm is used to solve non-linear least squares problems, combining the gradient descent method and the Gauss–Newton method (Lourakis 2005). All additional "outlier" matches are then removed.

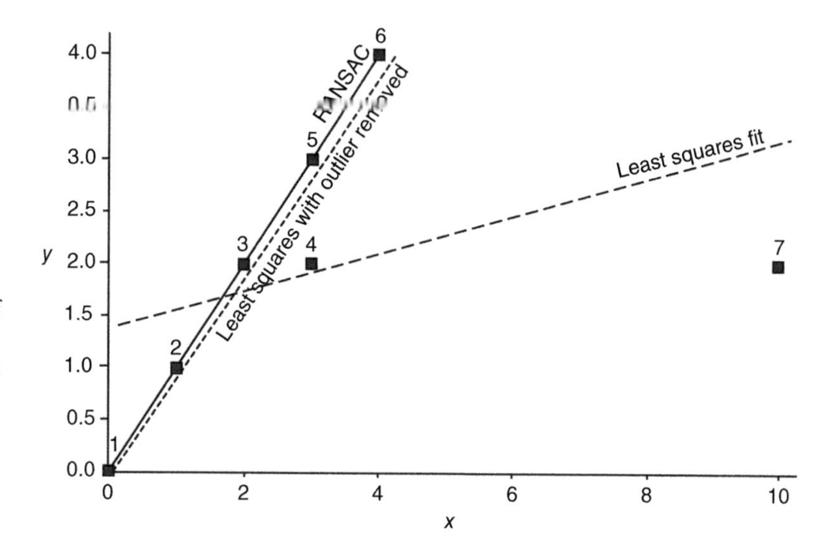


**Figure 3.7** Demonstration of the RANSAC algorithm by comparison with least squares fitting in the presence of an outlier (point 7). *x* and *y* are synthetic demonstration data sets. Source: Redrawn from Torr and Murray (1997, p. 272)

In the case that only a small number of "inlier" matches remain (<20 in Snavely et al. 2008), all matches are removed from consideration.

Other approaches to remove geometrically inconsistent keypoint matches are available and applied in different software. Indeed, RANSAC has been shown to perform poorly when outliers are much greater in number than inliers (owing to the random sampling inherent in the method) as they can distort a fitting process. Rousseeuw (1987) presented calculations for the minimum number of samples required to give a high probability (i.e. 95 %) that a good subsample is selected; this relies on the knowledge of the fraction of contaminated data. Where greater than 50% of data is contaminated, RANSAC performs poorly. Lowe (2004) notes that with keypoint correspondences from the SIFT algorithm inliers may contribute less than 1% of the total dataset. RANSAC is one of the many "hypothesise-and-test" frameworks (Nistér 2005); others are available. For example, Torr and Zisserman (2000) describe a similar algorithm of maximum likelihood estimation sample consensus (MLESAC) which improves on RANSAC by using the log likelihood of the fitted solution, that is, incorporating the error distribution, rather than simply the number of correspondences below a threshold. Lowe (2004) applies the Hough transform (Ballard & Brown 1982) to relate two images, where the parameter space is divided into cells, and each datum adds a vote to each cell of the parameter space which has parameters consistent with that datum. When the "votes" of each data point are accumulated, clusters of these votes can then be used to identify possible solutions. However, Lowe (2004) limits this technique to the determination of the best affine projection parameters as the high dimensionality of the fundamental matrix (seven parameters) would require much coarser quantisation of the parameter space to be feasible given the exponentially increased computational expense (Torr & Zisserman 2000).

With the keypoints limited to those with geometrically consistent matches, the links between every image pair can be identified and organised into tracks: connected sets of matching keypoints through the library of images used in the reconstruction (Snavely et al. 2008). A minimum of two keypoints located in three images is required for a track. Where the same keypoint occurs twice in a single image, the track is considered inconsistent. Maps of consistent tracks can then be made, identifying the connectivity of each image. These tracks are utilised in the steps that follow.

## 3.5 Structure from Motion

End users of SfM-MVS in the geosciences often refer to the entire process chain described here as SfM, but it is the single process of simultaneously estimating the 3D geometry (or structure) of a scene and the different camera poses (i.e. motion) that is more technically known as "SfM" (Ullman 1979). Using the geometrically correct feature correspondences identified in the previous section, SfM aims to reconstruct simultaneously: (i) 3D scene structure, (ii) camera positions and orientations (i.e. pose estimation or extrinsic calibration), and often (iii) intrinsic camera calibration parameters.

The extrinsic camera parameters of (ii) represent the rigid body transformation between the 3D scene coordinates and the camera coordinate system. Many different camera models exist to describe the intrinsic parameters of (iii). The most common model is a perspective projection described by a pinhole camera model; others include affine projections, orthographic projections, and push-broom models.

Intrinsic camera parameters are defined by a $3 \times 3$ upper triangular matrix known as the camera calibration matrix **K**:

$$\mathbf{K} = \begin{bmatrix} a_u & s & u_0 \\ 0 & a_v & v_0 \\ 0 & 0 & 1 \end{bmatrix}$$

where $\alpha_u$ and $\alpha_v$ scale the image in the $x$ and $y$ directions, respectively, and $s$ represents skew. Under the assumption of square pixels $s = 0$ and $\alpha_u = \alpha_v = \alpha$ where $\alpha$ is considered to be the focal length of the lens in units of the pixel dimension. The principal point $(u_0, v_0)$ is defined as the location on the image plane which intersects the optical axis.

Further intrinsic parameters are required to model internal aberrations (i.e. radial distortion parameters), assuming the cameras have not been pre-calibrated. Radial distortion causes image points to be displaced in a radial direction from the centre of distortion, which is often assumed to be the principal point, and can be corrected for with the knowledge of two coefficients of the distortion function ($k_1$ and $k_2$). The degree to which this radial distortion is incorporated into camera models is variable. MicMac (see Section 4.5) incorporates five coefficients of radial distortion (Oúedraogo et al. 2014), for example.

**Bundle adjustment** produces *jointly optimal* 3D structure and viewing parameter (pose and/or calibration) estimates (Granshaw 1980; Triggs et al. 2000). The words "jointly optimal" apply here as the parameter estimates that apply to both structure, and camera variations are made by minimising the value of a cost function that quantifies the model fitting error. Factorisation algorithms compute camera pose and 3D scene geometry using all images simultaneously using a singular value decomposition (e.g. Tomasi & Kanade 1992); however, these algorithms require all keypoints to be visible in all frames (although there are ways to deal with this limitation; Szeliski 2011). Sequential methods are a more popular alternative and are described next.

Parameter values must be assigned initial values before the non-linear parameter optimisation of the bundle adjustment. To avoid finding non-optimal (local minima) solutions to large-scale SfM problems, the scene reconstruction process typically begins with a single pair of images, referred to as the "initial pair." The initial pair should have a large number of matches and a large baseline (i.e. vastly different perspective) for a robust reconstruction. An initialisation of parameter values is required before scene reconstruction can begin. Where camera intrinsic parameters are initially unknown, a self-calibration method is required. Whilst Hartley and Zisserman (2003) note that self-calibration can be achieved with three or more frames or can recover focal lengths from two frames, thereby making assumptions about other camera parameters, Snavely et al. (2008) achieve this initialisation by limiting the choice of the initial pair to images for which focal length estimates are available (e.g. from exchangeable image file format or EXIF tags). They then obtain the remaining camera parameters using the five-point algorithm of Nistér (2004), which is a calibrated relative pose algorithm. Tracks visible in the initial pair are then triangulated to obtain initial estimates of feature positions (see Hartley & Sturm 1997).

The main goal of initialisation is to minimise the error between the projections of each track and the corresponding keypoints on the initial pair. Using this error as an optimality criterion to be minimised, the resulting non-linear least squares problem (Nocedal & Wright 2006) is solved using a two-frame bundle adjustment. Bundle adjustment originated in photogrammetry in the 1950s (Brown 1958; Slama 1980). A "bundle" refers to the bundles of light rays connecting camera centres to 3D points, and "adjustment" refers to the minimisation of the re-projection error (Szeliski 2011). Triggs et al. (2000) provide a detailed review of the bundle adjustment process.

With the re-projection error between the two images minimised, another camera is added into the optimisation (or multiple cameras). The camera containing the largest number of tracks whose 3D locations are already estimated is selected (or any camera with at least 75% of the maximum keypoint matches). Extrinsic camera parameters for the new camera are initialised using the direct linear transform technique (Abel-Aziz & Karara 1971), which uses a set of known control points (the existing known 3D locations) and maps the 2D coordinates of the new image onto the 3D coordinates of the 3D object space. This may be implemented within a RANSAC procedure (see Section 3.4) and also returns an upper triangular matrix **K** for initialisation values of the camera intrinsic parameters (along with EXIF tags). A further bundle adjustment step is run with this new image; however, only the new camera parameters and the points it observes are allowed to change. Where keypoints in the new image are observed by at least one other camera that has already been added into the model, all existing rays for that point are used to triangulate the point position. Where the maximum angle of separation between any pair of rays is less than a specified threshold (e.g. 2°), the new point is rejected.

To improve the accuracy of the solution, it is preferable to then perform a global bundle adjustment over all the cameras (Szeliski 2011) to refine the entire model. Minimisation of the cost function is an iterative process, fitting a local quadratic approximation to the cost function at each iteration (the Gauss–Newton approximation) or, where such models provide an inaccurate fit, using the gradient descent method (i.e. the LM algorithm mentioned previously). LM can converge quickly from a wide range of initialisations (Hiebert 1981). Yet with many cameras and multiple unknown parameters per camera, the bundle adjustment parameter space rapidly becomes high dimensional. Sparse bundle adjustment algorithms (Lourakis & Argyros 2009) reduce the otherwise intractable computational burden resulting from the high dimensionality of the problem by accounting for the lack of interaction among parameters for different cameras and 3D points.

Outlier tracks containing keypoints with a high re-projection error are removed after every run of the bundle adjustment optimisation. These can be defined according to the probability distribution of all re-projection errors, specific pixel error thresholds, or a combination of the two. Cameras are sequentially added to the model and the process mentioned before is repeated. The process ends when no remaining cameras contain a sufficient number of reconstructed 3D points to be reliably added to the model.

The SfM process produces a sparse point cloud and reconstructed camera poses. It is this sparse point cloud (once georeferenced) that is used in many applications of SfM, including those in the geosciences (e.g. Dandois & Ellis 2010; Fonstad et al. 2013); though further processing is required for more detailed higher-quality surface reconstructions (Rosnell & Honkavaara 2012). Most geoscience applications will apply MVS techniques to produce a much more dense point cloud.

## 3.6 Scale and Georeferencing

SfM-MVS only provides relative camera locations and scene geometry, so the point cloud output is generated in an arbitrary coordinate system. Absolute distances between cameras or between reconstructed points can never be recovered from images alone, regardless of how many cameras or points are used (Szeliski 2011). Georeferencing and scaling of the point cloud requires a minimum of three ground control points (GCPs) with *XYZ* coordinates for a seven-parameter linear similarity transformation, which comprises three global translation parameters, three rotation parameters, and one scaling parameter. Alternatively, "direct" georeferencing and scaling can be performed from known camera positions derived from real-time kinematic differential GPS (dGPS) measurements (Section 2.2.2) and an inertial measurement unit (Tsai et al. 2010; Turner et al. 2014). A common hybrid of the two georeferencing approaches uses direct georeferencing to provide approximate camera locations to initialise the bundle adjustment and then uses external GCPs to better constrain the solution (e.g. Ryan et al. 2015; Rippin et al. 2015).

dGPS or total station (TS) surveys of targets (Chapter 2) clearly visible in images typically provide the necessary real-world and absolute coordinates. Since identifying small features directly in point clouds can be difficult, many SfM-MVS software workflows allow the user to locate the target from the imagery directly (e.g. James & Robson 2012). The arbitrary coordinates of the targets from the SfM-MVS model are paired with the absolute coordinates of the GCPs and used to derive a similarity transformation. A larger number of targets than three is recommended, which is the absolute minimum required for a unique solution to the transformation. More details on collecting ground control data and other practical considerations are provided in Chapter 4.

## 3.7 Refinement of Parameter Values

Errors in the estimate of the intrinsic and extrinsic camera parameters arising from the SfM-MVS process can lead to non-linear deformations of the final model. The input of GCPs in the preceding step provides additional information on the 3D geometry that can be used to further refine camera parameters and reconstructed scene geometry. The known coordinates (and estimates of point error) provide an additional source of error in the minimisation of the non-linear cost function during the bundle adjustment step. With this external information included in the model, the bundle adjustment can be re-run to optimise the image alignment in light of this new information. Using the known reference coordinates supplied in the scaling and georeferencing step, some software packages (e.g. Agisoft PhotoScan) offer users the option of performing an additional optimisation of the image alignment whereby estimated internal camera parameters and 3D points are adjusted to minimise the sum of the re-projection error and the georeferencing error. This optimisation can improve survey accuracy by an order of magnitude, but as noted by Javernick et al. (2014), the transformation algorithms of Agisoft PhotoScan are not fully disclosed. The spatial distribution of GCPs is also crucial for this optimisation process, where GCPs do not adequately cover the area of interest and optimisation may be detrimental to the overall survey accuracy. Caution is therefore advised when undertaking this step.

## 3.8 Clustering for MVS

Before MVS techniques are applied to the point cloud there is an additional, optional step that may be required in projects with large image sets. Some MVS algorithms solve a depth map for each image in turn (using nearby images) and then merge the separate reconstructions (e.g. Micusik & Kosecka 2009). This permits parallelisation but at the expense of noisy and highly redundant depth maps that require further post-processing to clean and merge (Furukawa et al. 2010).

In contrast, many of the best-performing MVS algorithms reconstruct scene geometry globally using all images simultaneously (e.g. Pons et al. 2007). When the number of images increases, the computational burden of such an approach increases rapidly and issues of scalability emerge. Random access memory (RAM) requirements increase with the number of images used in the reconstruction and place a practical limit on the number of images that can be matched simultaneously.

The solution to this RAM problem is image clustering, that is, splitting a large project into chunks. Furukawa et al. (2010) detail a pre-processing step known as clustering views for MVS (CMVS), which is a method whereby the image set is decomposed into overlapping view clusters to enable dense MVS reconstructions to run on the clusters separately. The sparse point cloud generated from SfM is used to produce overlapping image clusters of a manageable size such that each 3D point is reconstructed by at least one cluster. The basic idea underlying the image clustering approach is demonstrated in Fig. 3.8.

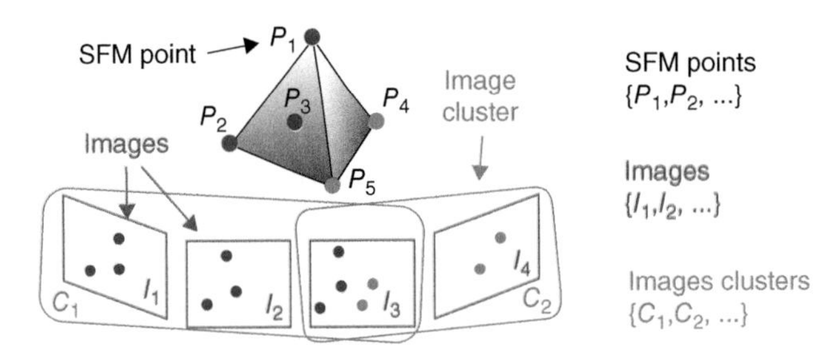


**Figure 3.8** The CMVS algorithm takes images $\{I_i\}$, SfM points $\{P_j\}$, and their associated visibility information $\{V_j\}$ to produce overlapping image clusters $\{C_k\}$. Source: Furukawa, Y., Curless, B., Seitz, S.M. & Szeliski, R. (2010) Towards internet-scale Multi-View Stereo. In: *IEEE Conference, Computer Vision and Pattern Recognition (CVPR)*, pp. 1434–1441. © IEEE. The image clustering algorithm described here is available at: http://www.di.ens.fr/cmvs/.

The total number of images in output clusters is minimised, and redundant images are removed from the reconstruction such that all SfM points within it are already well reconstructed (determined by camera baselines and pixel sampling rates) in at least one image cluster. A maximum image cluster size is specified such that each cluster is small enough for MVS reconstruction. Further point filters are applied in the cluster merging process after MVS algorithms have been applied as detailed in the following section, such that only relatively high-quality points are retained with an inter-cluster visibility consistency enforced (Furukawa et al. 2010).

Similarly, other software packages (e.g. Agisoft PhotoScan) permit users to manually identify "chunks" of an image set that are used to split up the MVS processing step and reduce memory requirements. The separate chunks are then aligned into a single point cloud.

## 3.9 MVS Image Matching Algorithms

A *sparse* point cloud generated by SfM is often only an intermediary step in the production of much more *dense* point clouds using MVS. The goal of MVS is to provide a complete 3D scene reconstruction from a collection of images of known camera intrinsic and extrinsic parameters. Compared with a sparse point cloud generated by SfM, a dense point cloud generated by MVS shows an increase in the point density of at least two orders of magnitude.

There is a wide variety of MVS algorithms (Seitz et al. 2006), and these can generally be divided into four classes: (i) *Voxel-based methods* represent the 3D scene volume directly using voxel occupancy grids (e.g. Seitz & Dyer 1999). These methods are relatively simple but are limited in accuracy by the resolution of the voxel grid, and they require knowledge of the bounding box that contains the scene. (ii) *Surface evolution-based methods* use deformable polygonal meshes that are iteratively evolved to minimise a cost function (e.g. Furukawa & Ponce 2009). These surface evolution-based algorithms require an initialisation (e.g. using a visual hull model) which limits their applicability, especially in large-scale scenes (Shen 2012). (iii) *Depth-map merging methods* compute individual depth maps for each image which are then combined into a single 3D model (e.g. Li et al. 2010). A depth map is an image representing the distance from the viewpoint to the 3D scene objects (Fig. 3.9). These algorithms avoid the need to resample on a 3D domain and are more flexible for crowded scenes. (iv) *Patch-based methods* represent scenes by collections of small patches (or surfels) (e.g. Lhuillier & Quan 2005), which are both simple and effective and do not require initialisation.

Furukawa and Ponce (2010) describe a **patch-based MVS** (PMVS) algorithm that is used widely and has performed well in tests comparing MVS algorithms (e.g. Ahmadabadian et al. 2013). The PMVS algorithm proceeds in three main steps, which due its widespread use are described briefly here: (i) matching features, (ii) expanding patches, and (iii) filtering incorrect matches.

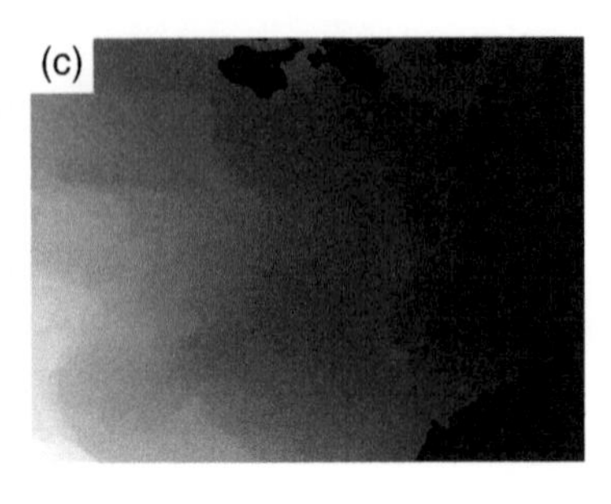


**Figure 3.9** (a) Image of an eroding river bank on the River Mersey as part of a detailed SfM-MVS survey, (b) normal vector calculated for that image, and (c) depth map for the same image. Black areas (typically vegetation or the water surface) have not been reconstructed. Source: Images were exported from Agisoft PhotoScan.

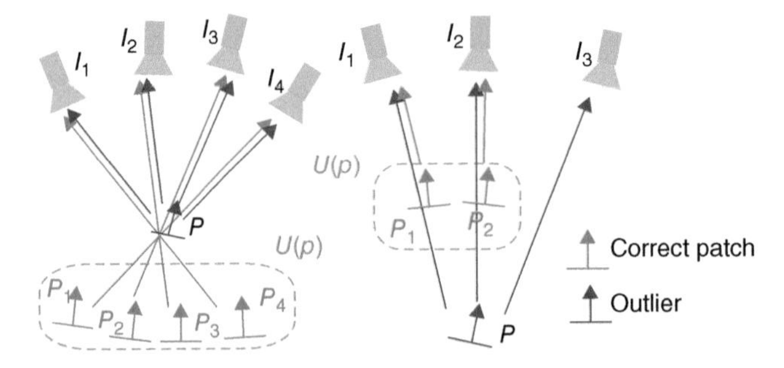


**Figure 3.10** Example of the visibility consistency filter used in PMVS. The filter enforces global visibility consistency to remove outliers (red patches). In both panels, *U(p)* denotes a set of patches that is inconsistent in visibility information with *p*. Source: Furukawa, Y. & Ponce, J. (2010) Accurate, dense and robust multiview stereopsis. *IEEE Transactions on Pattern Analysis and Machine Intelligence*, **32**, 1362–1376. © IEEE.

First, in the matching step, difference-of-Gaussian and Harris operators are used to detect corner and "blob" features, which are then matched across multiple images. Local photometric consistency is evaluated by normalised cross correlation (other photometric discrepancy functions are often used in this step). Owing to a lack of regularisation at the patch generation step, the dense reconstruction of PMVS relies on reliable texture information which may result in gaps in the dense point cloud on poor texture surfaces. Second, in the expansion step, starting with these initial matches (sparse patches), the neighbouring pixels in the images where the patch is projected are considered for expansion. This creates the dense patches by expanding the reconstruction. The expansion does not take place where neighbouring image cells are already reconstructed or where depth discontinuities occur. Third, in the filtering step, visibility constraints are then used to filter out incorrect matches by accounting for occlusion in the models. Global visibility consistency is enforced by filtering out outlier patches as demonstrated in Fig. 3.10. Further filters are applied to remove patches visible on only a small number of depth maps (Furukawa & Ponce 2010).

In PMVS, the expansion and filtering process is repeated several times (typically three). Other patch-based methods replace this iterative expansion and filtering process with greedy expansion procedures (e.g. Lhuillier & Quan 2005). In all cases, the result is a dense point cloud that has a similar point density to terrestrial laser scanner data.

## 3.10 Summary

This chapter has presented an overview of a typical workflow implemented in many SfM-MVS systems, from a set of images to a georeferenced, dense point cloud. It should be emphasised that the particular details will vary between software packages, but the vast majority of geoscience implementations share many features with the workflow:

- Feature detection
- Keypoint correspondence
- Identifying geometrically consistent matches
- Structure from Motion (SfM)
- Scale and georeferencing
- Optimisation of image alignment
- Clustering for MVS
- MVS image matching algorithms

Continuing developments in many of the processing steps (e.g. feature matching, bundle adjustment algorithms, and MVS algorithms) suggest that further refinements and improvements will be made as computer processing speeds increase, random memory requirements reduce, and as point densities and point accuracies increase. End users should be aware of the arbitrary parameters employed within the SfM-MVS workflow, which have a large effect on the data processing speed and the resultant point cloud density and accuracy. As the technology matures and practitioners demand a greater ability to adjust such parameters, it is likely that the situation will change in the future. At present, open-source code packages generally offer the user greater transparency as to the specific workflow implemented than commercially available software.

## References

Abel-Aziz, Y.I. & Karara, H.M. (1971) *Direct Linear Transformation from Comparator Coordinates Into Object Space Coordinates*, pp. 1–18. American Society of Photogrammetry, Urbana, IL.

Ahmadabadian, A.H., Robson, S., Boehm, J., Shortis, M., Wenzel, K. & Fritsch, D. (2013) A comparison of dense matching algorithms for scaled surface reconstruction using stereo camera rigs. *ISPRS Journal of Photogrammetry and Remote Sensing*, **78**, 157–167.

Arya, S., Mount, D.M., Netanyahu, N.S., Silverman, R. & Wu, A.Y. (1998) An optimal algorithm for approximate nearest neighbour searching in fixed dimensions. *Journal of the ACM*, **45**, 891–923.

Ballard, D.H. & Brown, C.M. (1982) *Computer Vision*. Prentice-Hall, Upper Saddle River, NJ.

Baumberg, A. (2000) Reliable feature matching across widely separated views. In: *Proceedings of the IEEE Conference on Computer Vision and Pattern Recognition*, Vol. **1**, June 2000, pp. 774–781. Hilton Head Island, SC.

Bay, H., Ess, A., Tuytelaars, T. & van Gool, L. (2008) Speeded-up robust features (SURF). *Computer Vision Image Understand*, **110**, 346–359, Similarity Matching in Computer Vision and Multimedia.

Beis, J.S. & Lowe, D.G. (1997) Shape indexing using approximate nearest-neighbour search in high-dimensional spaces. In: *Proceedings, IEEE Computer Society Conference, Computer Vision and Pattern Recognition*, 1997, pp. 1000–1006. San Juan.

Bentley, J.L. (1975) Multidimensional binary search trees used for associative searching. *Communications of the ACM*, **18** (**9**), 509–517.

Brown, D. (1958) A solution to the general problem of multiple station analytical stereo triangulation. Technical Report 43, RCA-MTP.

Bustos, B., Deussen, O., Hiller, S. & Keim, D.A. (2006) A graphics hardware accelerated algorithm for nearest neighbour search. *International Conference of Computational Science*, **4**, 196–199.

Calonder, M., Lepetit, V., Strecha, C. & Fua, P. (2010) BRIEF: binary robust independent elementary features. In: K. Daniilidis, P. Maragos & N. Paragios (eds), *Computer Vision—ECCV 2010, Lecture Notes in Computer Science*, **6314**, pp. 778–792. Springer, Berlin Heidelberg.

Dandois, J.P. & Ellis, E.C. (2010) Remote sensing of vegetation structure using computer vision. *Remote Sensing*, **2**, 1157–1176.

Fischler, M.A. & Bolles, R.C. (1981) Random sample consensus: a paradigm for model fitting with applications to image analysis and automated cartography. *Communications of the ACM*, **24** (**6**), 381–395.

Fonstad, M.A., Dietrich, J.T., Courville, B.C., Jensen, J.L. & Carbonneau, P.E. (2013) Topographic Structure from Motion: a new development in photogrammetric measurement. *Earth Surface Processes and Landforms*, **38**, 421–430.

Förstner, W. (1986) A feature-based correspondence algorithm for image matching. *International Archives Photogrammetry & Remote Sensing*, **26** (**3**), 150–166.

Friedman, J.H., Bentley, J.L. & Finkel, R.A. (1977) An algorithm for finding best matches in logarithmic expected time. *ACM Transactions on Mathematical Software*, **3**, 209–226.

Furukawa, Y. & Ponce, J. (2009) Carved visual hulls for image-based modelling. *International Journal of Computer Vision*, **81**, 53–67.

Furukawa, Y. & Ponce, J. (2010) Accurate, dense and robust multiview stereopsis. *IEEE Transactions on Pattern Analysis and Machine Intelligence*, **32**, 1362–1376.

Furukawa, Y., Curless, B., Seitz, S.M. & Szeliski, R. (2010) Towards internet-scale Multi-View Stereo. In: *IEEE Conference, Computer Vision and Pattern Recognition (CVPR)*, June 13–18, 2010, pp. 1434-1441. San Francisco, CA.

Granshaw, S.I. (1980) Bundle adjustment methods in engineering photogrammetry. *The Photogrammetric Record*, **10**, 181–207.

Harris, C. & Stephens, M.J. (1988) A combined corner and edge detector. In: *Alvey Vision Conference*, pp. 147–152.

Hartley, R.I. (1997) In defense of the eight-point algorithm. *IEEE Transactions on Pattern Analysis and Machine Intelligence*, **19**, 580–593.

Hartley, R.I. & Sturm, P. (1997) Triangulation. *Computer Vision and Understanding*, **68**, 146–157.

Hartley, R. & Zisserman, A. (2003) *Multiple View Geometry in Computer Vision.* Cambridge University Press, Cambridge.

Hiebert, K. (1981) An evaluation of mathematical software that solves nonlinear least squares problems. *ACM Transactions on Mathematical Software*, **7** (**1**), 1–16.

James, M.R. & Robson, S. (2012) Straightforward reconstruction of 3D surfaces and topography with a camera: accuracy and geoscience application. *Journal of Geophysical Research, Earth Surface*, **117**, F03017. doi: 10.1029/2011JF002289.

Javernick, L., Brasington, J. & Caruso, B. (2014) Modelling the topography of shallow braided rivers using Structure-from-Motion photogrammetry. *Geomorphology*. doi: 10.1016/j.geomorph.2014.01.006.

Levenberg, K. (1944) A method for the solution of certain non-linear problems in least squares. *The Quarterly of Applied Mathematics*, **2**, 164–168.

Lhuillier, M. & Quan, L. (2005) A quasi-dense approach to surface reconstruction from uncalibrated images. *IEEE Trans. Pattern Analysis and Machine Intelligence*, **27**, 418–433.

Li, J., Li, E., Chen, Y., Xu, L. & Zhang, Y. (2010) Bundled depth-map merging for Multi-View Stereo. In: *IEEE Conference, Computer Vision and Pattern Recognition (CVPR)*, June 13–18, 2010, pp. 2769–2776. San Francisco, CA.

Longuet-Higgins, H.C. (1981) A computer algorithm for reconstructing a scene from two projections. *Nature*, **293**, 133–135.

Lourakis, M.I.A. (2005) A brief description of the Levenberg-Marquardt algorithm implemented by levmar. *Foundation of Research and Technology*, **4**, 1–6.

Lourakis, M.I.A. & Argyros, A.A. (2009) SBA: a software package for generic sparse bundle adjustment. *ACM Transactions on Mathematical Software*, **36** (**1**article 2). doi: 10.1145/1486525.1486527.

Lowe, D.G. (1999) Object recognition from local scale-invariant features. In: *International Conference on Computer Vision*, September 20–27, 1999, pp. 1150–1157. Corfu, Greece.

Lowe, D.G. (2001) Local feature view clustering for 3D object recognition. In: *Proceedings of the IEEE Conference on Computer Vision and Pattern Recognition*, 2001, 7 pp. Kauai, Hawaii.

Lowe, D.G. (2004) Distinctive image features from scale-invariant keypoints. *International Journal of Computer Vision*, **60**, 91–110.

Lucas, B.D. & Kanade, T. (1981) An iterative image registration technique with an application in stereo vision. In: *International Joint Conference on Artificial Intelligence*, April 24–28, 1981, pp. 674–679. Vancouver, British Columbia.

Marquardt, D.W. (1963) An algorithm for least-squares estimation of nonlinear parameters. *Journal of the Society for Industrial and Applied Mathematics*, **11**, 431–441.

Matas, J., Chum, O., Urban, M. & Pajdla, T. (2004) Robust wide baseline stereo from maximally stable extremal regions. *Image and Vision Computing*, **22** (**10**), 761–767.

Micusik, B. & Kosecka, J. (2009) Piecewise planar city 3D modeling from street view panoramic sequences. In: *IEEE Conference, Computer Vision and Pattern Recognition. CVPR 2009*, June 20–25, 2009, pp. 2906–2912. Miami, FL.

Mikolajczyk, K. & Schmid, C. (2005) A performance evaluation of local descriptors. *IEEE Transactions on Pattern Analysis and Machine Intelligence*, **27**, 1615–1630.

Mikolajczyk, K., Tuytelaars, T., Schmid, C. et al. (2005) A comparison of affine region detectors. *International Journal of Computer Vision*, **65** (**1/2**), 43–72.

Moravec, H. (1983) The Stanford cart and the CMU rover. *Proceedings of the IEEE*, **71** (7), 872–884.

Morel, J. & Yu, G. (2009) ASIFT: a new framework for fully affine invariant image comparison. *SIAM Journal on Imaging Sciences*, **2**, 438–469.

Muja, M. & Lowe, D.G. (2009) Fast approximate nearest neighbours with automatic algorithm configuration. *VISAPP*, **1**, 331–340.

Nistér, D. (2004) An efficient solution to the five-point relative pose problem. *IEEE Transactions on Pattern Analysis and Machine Intelligence*, **26**, 756–777.

Nistér, D. (2005) Pre-emptive RANSAC for live Structure and Motion estimation. *Machine Vision and Applications*, **16** (**5**), 321–329.

Nocedal, J. & Wright, S.J. (2006) *Numerical Optimization*. Springer, New York.

Oúedraogo, M.M., Degré, A., Debouche, C. & Lisein, J. (2014) The evaluation of unmanned aerial systems-based photogrammetry and terrestrial laser scanning to generate DEMs of agricultural watersheds. *Geomorphology*. doi: 10.1016/j.geomorph.2014.02.016.

Pons, J.-P., Keriven, R. & Faugeras, O. (2007) Multi-View Stereo reconstruction and scene flow estimation with a global image-based matching score. *International Journal of Computer Vision*, **72**, 179–193.

Rippin, D.M., Pomfret, A. & King, N. (2015) High resolution mapping of supraglacial drainage pathways reveals link between micro-channel drainage density, surface roughness and surface reflectance. *Earth Surface Processes and Landforms*. doi: 10.1002/esp.3719.

Rosnell, T. & Honkavaara, E. (2012) Point cloud generation from aerial image data acquired by quadrocopter type micro unmanned aerial vehicle and a digital still camera. *Sensors*, **12**, 453–480.

Rousseeuw, P.J. (1987) *Robust Regression and Outlier Detection*. John Wiley & Sons, Inc., New York.

Ryan, J.C., Hubbard, A.L., Box, J.E. et al. (2015) UAV photogrammetry and Structure from Motion to assess calving dynamics at Store Glacier, a large outlet draining the Greenland ice sheet. *The Cryosphere*, **9** (**1**), 1–11.

Seitz, S.M. & Dyer, C. (1999) Photorealistic scene reconstruction by voxel coloring. *International Journal of Computer Vision*, **35** (**2**), 151–173.

Seitz, S.M., Curless, B., Diebel, J., Scharstein, D. & Szeliski, R. (2006) A comparison and evaluation of Multi-View Stereo reconstruction algorithms. In: *IEEE Computer Society Conference, Computer Vision and Pattern Recognition*, Vol. **1**, June 17–22, 2006, pp. 519–528.

Shen, S. (2012) Depth-map merging for Multi-View Stereo with high resolution images. In: *Pattern Recognition (ICPR), 21st International Conference*, November 11–15, 2012, pp. 788–791. Tsukuba.

Slama, C. (1980) *Manual of Photogrammetry*. American Society of Photogrammetry, Falls Church, VA.

Snavely N. (2008) Scene reconstruction and visualization from internet photo collections, PhD thesis, University of Washington.

Snavely, N., Seitz, S.N. & Szeliski, R. (2008) Modeling the world from internet photo collections. *International Journal of Computer Vision*, **80**, 189–210.

Strecha, C., Bronstein, A., Bronstein, M. & Fua, P. (2012) LDAHash: improved matching with smaller descriptors. *IEEE Transactions on Pattern Analysis and Machine Intelligence*, **34**, 66–78.

Sunglok, C., Taemin, K. & Wonpil, Y. (2009) Performance evaluation of RANSAC family. In: *Proceedings of the British Machine Vision Conference (BMVC'09)*, September 7–10, 2009. London.

Szeliski, R. (2011) *Computer Vision: Algorithms and Applications*. Springer, London.

Tomasi, C. & Kanade, T. (1992) Shape and motion from image streams under orthography: a factorization method. *International Journal of Computer Vision*, **9**, 137–154.

Torr, P.H.S. & Murray, D.W. (1997) The development and comparison of robust methods for estimating the fundamental matrix. *International Journal of Computer Vision*, **24**, 271–300.

Torr, P.H.S. & Zisserman, A. (2000) MLESAC: a new robust estimator with application to estimating image geometry. *Computer Vision and Image Understanding*, **78**, 138–156.

Triggs, B., McLauchlan, P.F., Hartley, R.I. & Fitzgibbon, A.W. (2000) Bundle adjustment – a modern synthesis. In: B. Triggs, A. Zisserman & R. Szeliski (eds), *Vision Algorithms '99, LNCS 1883*, pp. 298–372. Springer-Verlag, Berlin Heidelberg.

Tsai, M.L., Chiang, K.W., Huang, Y.W., et al. (2010) The development of a direct georeferencing ready UAV based photogrammetry platform. In: *Proceedings of the 2010 Canadian Geomatics Conference and Symposium of Commission I*, June 15–18, 2010. Calgary, AB.

Turner, D., Lucieer, A. & Wallace, L. (2014) Direct georeferencing of ultrahigh-resolution UAV imagery. *IEEE Transactions on Geoscience and Remote Sensing*, **52**, 2738–2745.

Ullman, S. (1979) The interpretation of Structure from Motion. *Proceedings of the Royal Society B*, **203**, 405–426.

Zhang, Z. (1996) Determining the epipolar geometry and its uncertainty: a review. Technical Report RR-2927, INRIA.

## Further Reading/Resources

Details of the SIFT object recognition system are provided in Lowe (2004). The SIFT code is available from http://www.cs.ubc.ca/~lowe/keypoints/. A detailed SfM workflow for use on internet photo collections is described in Snavely et al. (2008). Triggs et al. (2000) provide details on bundle adjustment algorithms. A relatively accessible description of computer vision techniques more generally is found in Szeliski (2011).

# 4 Structure from Motion in Practice

Abstract
Advances in computing hardware and software have reduced the financial and computational expense of generating topographic data using Structure from Motion - Multi-View Stereo (SfM-MVS), making it accessible to enthusiasts and experts alike. Here, we review the most commonly used platforms for acquiring input images and compare them based on expense and practicality. Whilst helicopter or light aircraft overflights offer the elevation needed for wide image swaths, the cost is prohibitive for many users. In contrast, ground-based, hand-held photography can be cost-free and has the advantage of providing very fine-resolution imagery. The quality of derived products is less dependent on the sensor used, with camera phones providing sufficient resolution for many applications, but more dependent on the geometry of camera positions and the distribution of external ground control. Clusters of images with short spatial baselines can degrade the quality of derived data, and ground control that is either insufficient or inappropriately distributed can introduce significant external error into the final model. Most software will report these errors, but many are black box, giving the user very limited information to assess the final model quality, or indeed any control over the processing parameters. This is true of most commercial software; in contrast, a host of open-source applications are available that allow the user to modify processing parameters and build on existing code, but they tend to be less user-friendly and generally may require some programming expertise. There is therefore an effective trade-off between accessibility and quality control, the latter being an issue that is currently given insufficient attention in the literature.

Keywords
platform; sensor; software; ground control; filter; survey planning

*Structure from Motion in the Geosciences*, First Edition. Jonathan L. Carrivick, Mark W. Smith, and Duncan J. Quincey.

Companion Website: www.wiley.com/go/carrivick/structuremotiongeosciences

## 4.1 Introduction

Compared to other digital topographic surveying methods as overviewed in Chapter 2, image-based approaches have dramatically decreased in cost and increased in ease of use in recent years primarily because of advances in technology. Specifically the production of lightweight components for the mobile telephone industry has led to remotely operated airborne platforms (for image acquisition) becoming a viable investment for many users. In parallel, the development of software that can make use of increased availability of computation power has led to the possibility to handle hundreds of images (and their matching) simultaneously. Geoscientists can therefore generate three-dimensional (3D) data comparable in scale and accuracy to that either from terrestrial laser scanning or airborne light detection and ranging (LiDAR) or from classic photogrammetric approaches, simply by using their standard desktop PC and a hand-held camera. In addition, the automation of the Structure from Motion–Multi-View Stereo or SfM-MVS workflow (see Chapter 3) makes the production of topographic data an easy process, with limited expertise required to operate most off-the-shelf SfM-MVS software, giving enthusiasts and experts the same opportunity to make their own 3D data sets. This democratisation makes the potential applications of SfM-MVS in the geosciences almost boundless.

One of the main strengths of the SfM-MVS approach is its flexibility in the type, number, scale, and positioning of input images that it can handle in the workflow. Numerous combinations of platforms, sensors, and software have been employed to achieve the same end goal, and development in some areas has outpaced developments in others. From a practical point of view, the user has to weigh up a number of factors in making these choices: cost, accessibility (and portability), experience, and fitness for purpose (i.e. resolution and coverage). This can be confusing for the novice, and even many experts are not aware of the diversity of the available options. It is also not always clear to SfM-MVS users how important the quality of the input images is to the accuracy of the final model. Simple and practical steps to ensure images and external ground control points (GCPs) are suitably distributed across the scene of interest and can go a long way towards reducing errors to a level offered by more traditional approaches. Clear advice on how to do this is sadly lacking in the literature though.

The aim of this chapter is thus to provide an overview of platforms, sensors, and software available to SfM-MVS users and, where appropriate, to assess their advantages and disadvantages. An additional and important focus is on the logistical aspects of acquiring imagery (camera geometry, lighting conditions, etc.) and how to ensure that external ground control is both visible and appropriately distributed within the reconstructed scene. Post-processing methods to remove spurious data points are reviewed, and methods for decimating the often unwieldy point cloud data into something more manageable are appraised. The chapter concludes with an assessment of the outstanding key issues in the acquisition and processing of imagery in the SfM-MVS workflow.

## 4.2 Platforms

Imagery for SfM-MVS can be acquired from almost any platform and may range from ground-based (hand-held or pole-mounted) through to airborne approaches (Fig. 4.1). As is often the case, the simpler methods often produce some of the best results. From the viewpoint of cost-effectiveness, ground-based approaches are clearly preferable. In addition, the close range of ground-based approaches results in fine spatial resolution imagery and offers complete control over image acquisition. Of course, these advantages are offset by the limited spatial coverage one can achieve, so sometimes there is no alternative but to mount the sensor on a remote platform. In this case, a range of options exist (Table 4.1).

**Figure 4.1** Examples of platform types from which imagery may be acquired for SfM-MVS: (a) kite, (b) quadcopter, (c) pole/mast, (d) lighter-than-air blimp, and (e) gyrocopter.

**Table 4.1** A summary of the key platforms for acquiring SfM-MVS imagery with the main associated advantages and disadvantages, and key references demonstrating their use.

| Platform | Payload | Key advantages | Key disadvantages | Approximate cost (GBP) | References |
|---|---|---|---|---|---|
| Ground-based (hand-held) | Effectively unlimited | Cost; full control over image frequency/position; fine image resolution | Limited image swath | No cost | Bemis et al. (2014) and James and Robson (2012) |
| Mast/pole | 1–3 kg | Portability; cost; full control over image frequency/position | Poor stability in adverse weather; limited image swath | 10–500 | Mathews and Jensen (2012) and Plets et al. (2012) |
| Blimp | 3–5 kg | Cost; low maintenance; unlimited flying time; wide swath possible | Poor portability (steel canister); unstable in adverse weather | 500–5000 | Vericat et al. (2009) and Fonstad et al. (2013) |
| Fixed-wing UAV/multicopter | 5–10 kg | High maintenance costs; flying expertise required; pre-planned flight lines possible | Generally short battery life (flying time); high set-up cost for professional-grade systems | 5,000–25,000 | Bendig et al. (2012) and Dunford et al. (2009) |
| Kite | 3–5 kg | Portability; cost; low maintenance; unlimited flying time; can be deployed at high-elevation sites | Irregular winds can prevent flying; limited control over image frequency/position | 10–1000 | Smith et al. (2009) and Westoby et al. (2015) |
| Heli/gyrocopter or light aircraft | Effectively unlimited | Wide swath imagery; full control over image acquisition | Cost; flying not possible in adverse weather | 250–10,000 (flying time only) | James and Varley (2012) and Javernick et al. (2014) |

### 4.2.1 *Mast, Pole, or Boom*

Traditionally, low-cost aerial photography has been acquired by a sensor mounted atop a mast, pole, or boom; yet this remains one of the most viable platforms from which SfM-MVS imagery can be acquired. It has clear advantages in terms of cost efficiency (the simplest set-up could cost as little as GBP10 to GBP20 if on a shoestring budget, or up to several hundred pounds for a bespoke, extendable, fibreglass boom, for example, http://bit.ly/1HWlrJZ), and the user can retain full control over the frequency and target of image acquisition with relative ease. Remote triggering and wireless connections to smartphones facilitate pole-based image acquisition. Modern, extendable poles tend to be manufactured using lightweight materials and are thus very portable. In windy conditions retaining stability can be challenging, but certainly not as challenging as retaining stability with an airborne platform. On the downside, poles, masts, and booms are generally limited by a moderate maximum operation height of 20 m, equating to an on-the-ground image swath of approximately 50 m to 60 m, making wide-area surveys (e.g. a several kilometre river reach) unfeasible.

### 4.2.2 *Unmanned Aerial Vehicles*

Rapid advances in microchip, Global Positioning System (GPS), and inertial motion unit (IMU) technology have seen a proliferation of relatively low-cost airborne platforms onto the market in the past 5 years. Many different terms are used to describe these platforms – the most common being **unmanned aerial vehicles** (UAVs), unmanned aerial systems (UAS), remotely piloted vehicles (RPVs), or simply drones. The terms "UAV," "RPV," and "drone" tend to refer to the aircraft itself, whereas UAS more commonly describes the entire set-up, including the remote control unit and the wireless data link. Here, we shall use the term UAV to describe all remote airborne platforms being operated or controlled by a ground-based user.

Given the recent surge in UAV operators, many countries have introduced regulations that govern when and where UAVs can be flown (Table 4.2).

There are four main types of UAV (Fig. 4.2):

1. **Self-propelled fixed-wing aircraft.** These are highly efficient (energy wise) and have long-range capability, facilitating large area surveys. Given their size and rigidity, they tend to be one of the most stable acquisition platforms. However, their large size (relative to other approaches) often demands a take-off and landing strip, which may not be feasible in remote and/or rugged terrain.

**Table 4.2** Current legislation governing non-commercial UAV activity in selected countries.

| | Restrictions | Further information |
|---|---|---|
| Canada | Model aircraft must be below 35 kg in weight, individually owned, and not profit seeking. Recommended rules of keeping aircraft in sight, not to fly close to airports or populated areas, or higher than 90 m | http://www.maac.ca/en/ (Model Aeronautical Association of Canada) http://www.tc.gc.ca/eng/menu.htm (Transport Canada) |
| Australia | Model aircraft flown "for sport, recreation, and education" are not regulated | http://www.casa.gov.au/ (Civil Aviation Safety Authority) |
| Mainland Europe | Under the jurisdiction of the European Aviation Safety Agency (EASA); need certification to fly | https://www.easa.europa.eu/ (search for E. Y013-01 for certification guidance) |
| UK | Model aircraft must be below 20 kg in weight; technically a "permit to fly" classification is required. Aircraft must be kept within visual line of sight (maximum 500 m horizontally; 400 ft vertically) | http://www.caa.co.uk/ (Civil Aviation Authority – search for CAP 722 for guidance document) |
| New Zealand | No limitations up to 25 kg; no operations permitted within 4 km of airports or above 400 ft; line of sight required at all times | http://www.caa.govt.nz/rpas/ (Civil Aviation Authority of New Zealand) |
| Brazil | No restriction to civilian use | None |
| Mexico | No restriction to civilian use | None |
| USA | Legislation currently in development; airspace authorisation is given at or below 200 ft; craft must be in line of sight at all times; exclusion zone of 5 miles around airports; weight limit of 25 kg | https://www.faa.gov/uas/model_aircraft/ (Federal Aviation Administration) |
| India | Legislation currently in development; approval for UAV flights require approval from the Air Navigation Service | http://dgca.nic.in/ (Directorate General of Civil Aviation) |

**Figure 4.2** Many types of fixed-wing, vertical take-off and landing, and multicopter platforms exist. Source: (a) Reproduced with permission from Colomina and Molina (2014); (b) Reproduced with permission from David Rippin, University of York; (c) Reproduced with permission from Stanford University http://robots.stanford.edu; and (d) Reproduced with permission from Duncan Quincey.

2. **Vertical take-off and landing dual rotor systems (e.g. heli).** These are highly flexible systems that can be deployed in almost any terrain. They have medium range (restricted in most cases by battery life), cannot be deployed in blustery conditions if stability is required, and given the energy required to lift and maintain the platform at high elevation, flying time is usually short.
3. **Airships.** These relatively large craft are not frequently deployed for capturing SfM-MVS imagery because of their high running costs, though their flight time and range are both long, making wide-area surveys possible.
4. **Multicopters.** Most flying enthusiasts own a multicopter because of their ease of operation and high flexibility and stability in most weather conditions. With time and technological development both purchase and maintenance costs are decreasing, and entire online communities are devoted to supporting multicopter developments and innovations. On the downside, these aircraft have limited range and flight time – typical battery life may be of the order of approximately 20 minutes. Multicopters are probably the most popular platform for collecting SfM-MVS imagery and have aided the production of the largest SfM-MVS 3D models to date (e.g. University

> of Maryland, Baltimore County (UMBC) Aerial Ecosynth). Many enthusiasts will endeavour at some point to build their own multicopter. That said, for the novice, there are many off-the-shelf aircraft that have guaranteed stability and functionality, which are crucial perhaps to avoid sparking a national security alert (Fig. 4.3).

The frames of multicopters are usually constructed of lightweight materials such as carbon fibre–reinforced plastic or aluminium. They can reach sizes of up to a metre and weigh (without payload) up to several kilograms. Many multicopters may take a payload of equal to this; some of the larger systems may take a payload of between 5 and 10 kg (Lejot et al. 2007). Most modern multicopters are equipped with precision sensors to finely tune rotor speeds, maintain a pre-programmed flightline in adverse conditions, sense low-battery levels, and return to a point of safety when necessary. Almost all UAVs need an IMU, which measures the relative state of the mobile unit with respect to an inertial reference frame (i.e. orientation, velocity, and position; Chao et al. 2010), and this component can account for a large part of the UAV set-up cost. Many commercial off-the-shelf products exist, but given their relative expense, many developers choose to build their own systems with low-cost inertial measurement units (IMUs). The drive to develop

NEWS US & CANADA

Home World UK England N. Ireland Scotland Wales Business Politics Health Education Sc

Africa Asia Australia Europe Latin America Middle East US & Canada

26 January 2015 Last updated at 18:42

709 Share

Flying 'quad copter' triggers White House alert

A White House spokesman said the device posed no threat.

**Figure 4.3** Novice multicopter pilots may want to choose their training locations carefully (see the website for the videos).

low-cost and light gyroscopes (that detect rotational attributes such as pitch, roll, and yaw) has been partly due to their inclusion as a standard feature in smartphones. Most IMUs will also include one or more accelerometers, and a magnetometer to aid in correcting orientation drift. For developers interested in building their own IMU, a quick Google search will yield numerous online tutorials and help pages.

Other hardware components often include brushless electrical motors, long carbon fibre propellors, and crucially a flight control unit (FCU). The FCU sends commands to the rotors and rotational speed is adjusted accordingly. The FCU is usually coupled to a differential GPS (dGPS) receiver and compass for holding position, returning home, and flying along waypoints. Waypoint routes are normally uploaded via external software (Fig. 4.4). The information required for each route commonly includes the global position of each waypoint, the required flying height, the speed and heading of the craft as it approaches the waypoint, and any programming for external controls of the camera trigger. In some cases, a user may choose to deploy a pre-defined camera trigger (i.e. use a camera with an intervalometer) or use

**Figure 4.4** Pre-planning your flight with open-source software such as 3DR's Tower (https://3dr.com/software/) can ensure the area of interest is adequately covered. Source: Reproduced with permission.

an infrared trigger. Camera mountings vary widely, but most will comprise at least a couple of servo motors controlling pitch and roll to ensure nadir imaging.

### *4.2.3 Kites*

Kite aerial photography has been practised for several decades and has found considerable application in archaeology (Verhoeven et al. 2009) as well as occasional use in periglacial and proglacial geomorphology (Boike & Yoshikawa 2003; Westoby et al. 2014), for example. Their key advantages are their portability, low cost, low maintenance, and almost unlimited flying time, providing weather conditions are suitable. They can normally carry a payload of several kilograms, sufficient for a digital single-lens reflex (DSLR) camera, and given a sufficiently long tether, they can achieve much wider-swath imagery than a multicopter might. They are also practical for flying at high-elevation sites, where motorised platforms often fail because of low air density and where portability is a critical factor in platform choice. The reliance of kites on wind power alone is often problematic, however, as irregular winds are not suited for kite flying, and the size of the kite that can be flown is directly dependent upon the wind speed.

### *4.2.4 Lighter-than-Air Balloons*

Lighter-than-air balloons, often referred to as blimps, provide a complementary option to kite-based platforms. Where kites do not perform well in irregular or light winds, balloons and blimps can be used in both entirely windless and very light wind conditions. Indeed, many balloon systems feature a kite that sits beneath the balloon itself, partly to provide lift and partly to maximise the platform stability. They tend to be flexible in their set-up and are highly portable, providing some means of transporting the often cumbersome lighter-than-air canister are available. However, given their (low) weight, balloons become difficult to position and hold steady if the wind speed exceeds approximately 15 km $h^{-1}$ (Verhoeven et al. 2009). Payloads in excess of 300 kg have been reported for large balloon-based systems (e.g. Vierling et al. 2006), and flying heights in excess of 500 m are not uncommon.

Stationary tethered balloons have been used for some time (e.g. Vetrella et al. 1977; Church et al. 1998), but recently there has been increased use of more mobile platforms that provide an equally low-cost solution (Vericat et al. 2009). Their major limitation is their payload capability; small, easily portable, and low-cost balloons are typically constrained to operating loads of between 0.25 and 0.5 kg per 1 $m^3$ of lighter-than-air gas. For this reason, coupled dGPS and associated navigation systems are rarely deployed on balloon platforms, and low-grade uncalibrated compact cameras are usually the default sensor of choice. There is thus limited opportunity to acquire real-time camera position and attitude data, making exterior orientation estimation an integral part of the initial bundle adjustment (see Section 3.5), and the overall model quality highly dependent on the accuracy of external ground control. In addition, helium, or a low-density equivalent gas, is not always readily available or portable given its normal supply in a heavy steel canister.

### 4.2.5 *Aerial Photography*

Whilst most airborne platforms used to acquire SfM-MVS imagery are operated remotely, where sufficient funding exists the commissioning of a bespoke and manned overflight can provide significant time savings because (i) the maximum flying height of the aircraft is almost unrestricted, so image swath can be wide in comparison to remotely operated platforms and (ii) the sensor operator can simultaneously review the acquired imagery and communicate with the pilot to identify areas of deficient coverage, ensuring high-quality data are captured in a single flight mission. Data acquired for SfM-MVS on manned aircraft are normally via helicopter, with the camera operator clipped into the helicopter body but able to extend out of an open side panel, or more preferably, with the camera mounted on the heli skid and remotely activated, or on intervalometer control. Gyrocopters (or AutoGiros) offer a lower-cost alternative to a helicopter charter; whilst offering similar advantages to helicopter missions, the often open cockpit also increases the ease of image acquisition (see Box 5.1).

## 4.3 Sensors

The images required for SfM-MVS can be acquired from almost any sensor (Table 4.3). At one end of the image quality spectrum lies stills captured from online video footage of a feature or landscape, and multiple images (acquired by multiple sensors) stored on, for example, Internet photo sharing sites (Snavely et al. 2006). At the other end lies images acquired by professional-grade DSLR cameras or bespoke sensors for aerial reconnaissance (i.e. digital frame cameras). Such top-end sensors could even extend to those acquiring imagery at thermal or near-infrared wavelengths. In between, there is a common middle ground, where off-the-shelf compact cameras (Eisenbeiss & Zhand 2006), smartphone cameras (Klein & Murray 2009), and trail cameras dominate, and where the majority of SfM-MVS users find themselves not only because of ease and cost-effectiveness but also because the quality of the derived 3D products can often compete with those acquired from more expensive sensors (Thoeni et al. 2014).

**Table 4.3** A selection of sensors appropriate for SfM-MVS with their associated cost and technical specifications.

| Sensor | Effective pixels (Mp) | Resolution (pixels) | 35 mm focal length (equiv) | Sensor type | Weight | Cost (£) |
|---|---|---|---|---|---|---|
| iPhone 6 Plus | 7.99 | 2449 × 3264 | 29 mm | CMOS | 172 g | ~600 |
| Panasonic Lumix LX5 | 9.52 | 2520 × 3776 | 24–90 mm | CCD | 271 g | ~280 |
| Panasonic Lumix ZS20 | 14.1 | 3240 × 4230 | 24–480 mm | CMOS | 204 g | ~200 |
| Canon EOS 7D | 17.92 | 2345 × 5184 | Standard 29–216 mm (but lens dependent) | CMOS | 820 g (body) | ~500 |
| Acorn Trailcam 5310 | 5.0–12.0 | 4000 × 3000 | 6 mm | CMOS | 245 g | ~150 |
| Nikon 750 | 24.3 | 6016 × 4016 | Standard 24–120 mm (but lens dependent) | CMOS | 750 g | ~1500 |

Adapted from Thoeni et al. (2014).

The ultimate goal is to achieve many well-exposed photographs of the feature of interest at sufficient resolution for the matching algorithm to be able to perform effectively. The optimal resolution is therefore determined by the viewing geometry (distance from the feature), the size of the feature (in relation to that distance), the lighting conditions, and the number of images that will be acquired (many images at poor(er) resolution may be preferred to fewer images at fine(r) resolution). Of course, image resolution, contrast, and pixel sharpness tend to improve with the cost of the camera or sensor, yet many of the enhanced features of a top-end DSLR (e.g. imaging resolution ~30–40 MP) become redundant if large file sizes extend processing times beyond an acceptable level. In such cases, many users choose to degrade the resolution of their images prior to processing, realising that a large saving in processing time may lead to only a slight degradation in the quality of point cloud. Furthermore, many users also consider cheaper sensor options over more expensive units because of benefits relating to robustness and battery life, which become particularly important when working in remote or extreme environments.

From a practical point of view, a user needs to consider the platform on which the camera will be mounted. Clearly if imagery is to be collected from the ground (i.e. hand-held), the size and weight of the sensor is of low importance. However, if the sensor is to be mounted on a UAV, balloon, or kite, payload becomes a serious issue. Many small UAVs, balloons, and kites are limited in their ability to stabilise the sensor system or to take advantage of direct sensor georeferencing because of payload limitations (Nebiker et al. 2008); as a rule of thumb, most airborne platforms may only carry around 20–30% of the total weight of the system (often roughly 200–300 g). Activation of the sensor also needs to be an integral part of any airborne campaign. Only a handful of off-the-shelf cameras are equipped with an intervalometer, so many users have resorted to a remotely triggered or pre-programmed mechanical operation (i.e. with the shutter depressed by a pencil rubber mounted on a mechanical arm, or similar), a tethered or wireless ground-based operating system, or a smartphone app. It is also possible to rewrite the firmware operating the camera (see, e.g. Neitzel and Klonowski 2011).

To minimise power expenditure and also cost, light-weight consumer-grade digital cameras are thus most commonly used. At low elevation, most off-the-shelf units are able to achieve a ground resolution of centi- to decimetres and have sufficient stability (i.e. known and consistent interior orientation) to yield high-quality images suitable for point cloud derivation (Shortis et al. 2006). Most SfM-MVS algorithms assume that the camera is equipped with either a mechanical shutter or a global shutter CCD sensor, ensuring the frame is acquired in a single timestep. This is true of the majority of DSLRs and older compact cameras, but there is an increasing number of newer cameras equipped with complementary metal–oxide–semiconductor (CMOS) sensors, which scan the image row by row (Petrie & Walker 2007). These are also known as rolling shutter cameras. Roller shutters yield a reduction in power consumption and causes an improved data storage, but in the case of moving images they can lead to geometric distortion and so such sensors should be used with caution. Camera calibration can also be challenging because of the image stabilisation and automatic focussing algorithms employed within both "traditional" CCD and contemporary CMOS systems (Nebiker et al. 2008).

DSLR and fixed-focal-length cameras offer the most appropriate imagery for SfM-MVS because of their high image quality (James & Robson 2012). The internal geometry (or camera model) is most easily and commonly replicated for wide-angle lenses (equating to around 35 mm on a traditional SLR), whereas those with longer lenses (i.e. around three or more times the diagonal distance across the sensor) or even fish-eye lenses require bespoke models or algorithms (Micusik & Pajdla 2006). For this reason, although they are one of the most popular sensors for data capture, GoPro cameras have been shown to produce comparatively poor 3D data (Thoeni et al. 2014). That said, comparisons between point clouds generated from a range of sensors have shown that expensive DSLR cameras do not guarantee high-quality results either, with off-the-shelf compact cameras outperforming more expensive units in a number of cases (e.g. Thoeni et al. 2014). Our own analyses with very inexpensive (~£100) trail cameras (Fig. 4.5) have shown that an array of lower resolution but well-positioned cameras can produce comparable data to a series of high-quality images captured by a moving camera (see Box 7.1); the quality of the data capture strategy can therefore be as important as the quality of the imaging sensor (Micheletti et al. 2014).

## 4.4 Acquiring Images and Control Data

As has become clear in previous chapters, SfM-MVS depends entirely on the input of images taken from many viewpoints, and a common mistake to make is to take many pictures from the same location. For close-range SfM-MVS, pictures taken at each step by moving around the object of interest are ideal

**Figure 4.5** One time-lapse trail camera (out of an array of 15) deployed in West Greenland to derive multi-temporal point clouds of a lake-terminating ice cliff. See Box 7.1 for more information on this array.

(Fig. 4.6). For far-range SfM-MVS (e.g. when deriving models of the physical environment), pictures should be acquired around the landscape or catchment of interest to provide a 360° coverage if possible. Images taken around a horseshoe ridge line of the catchment can provide excellent data, for example, though this can be logistically difficult on occasions (e.g. when surveying a glacier where the ridgeline may be at 8000 m.a.s.l!). In such cases, the user may be restricted to finding several high-elevation but well-spaced viewpoints, or deploying an aircraft of some description. At large scales, and when working in the physical environment, recent work has shown that three main parameters affect the quality of the 3D model (Bemis et al. 2014):

1. Lighting conditions – glare from reflective surfaces and variable contrast across a scene can negatively affect point matching.
2. Changes in shadow length and surface albedo as a result of solar positioning also negatively impact on feature matching. Surveys should ideally be completed in less than 30 minutes and continued in following days if similar weather conditions prevail. Likewise, the camera operator should avoid casting a shadow over the study area.
3. Limited and/or poorly distributed stations result in model distortions and/or areas of missing data.

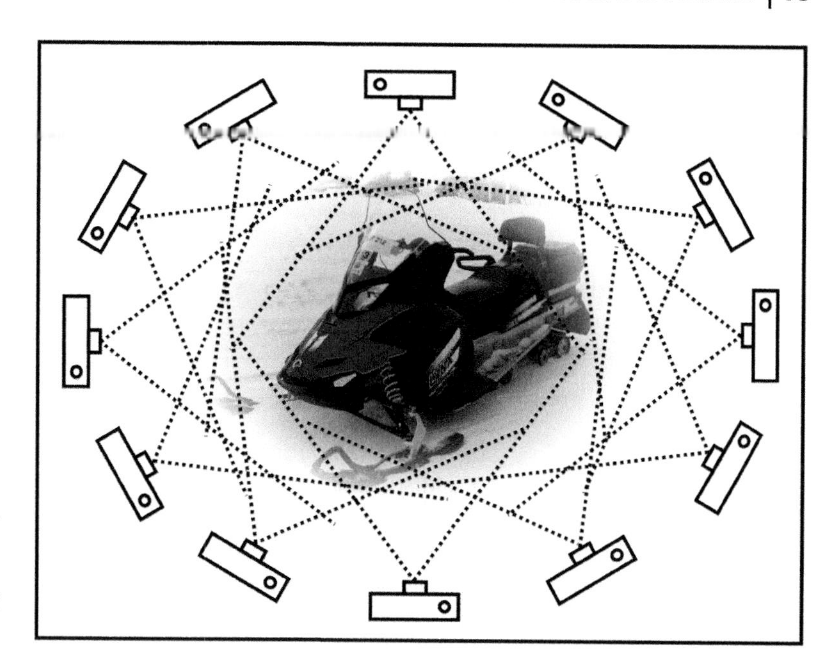

**Figure 4.6** Ideally, images should be acquired from many discrete locations, giving a full 360° coverage of the scene or object of interest. Features need to be visible in at least three images for them to be used in the formation of tracks (see Section 3.4).

Planning your SfM-MVS survey is therefore critical, and where possible images should be acquired such that the lighting is constant, the object of interest is fixed, and moving shadows and camera flash are minimised. Whilst some of these issues can be overcome by the use of a shorter focal-length camera (i.e. to reduce the number of images, and thus time, required), there is an associated reduction in the resolution of the generated model. On the other hand, whilst larger focal lengths can yield very fine resolution models, they can increase processing time non-linearly (e.g. as a general rule, doubling the number of photographs in a model can increase the computational time by a factor of four). In addition, the texture of the object or scene is important, and in particular the homogeneity of the colour composition. SfM-MVS depends on these textural differences to match features; if the object or scene is relatively devoid of features, the scale-invariant feature transform (SIFT) algorithm will fail (see Section 3.2).

It is difficult to arrive at a recommended number of images for each model, because this will differ depending on the camera parameters, the complexity of the scene, and the computational power available. It is often not the case that a higher number images equate to a higher accuracy reconstruction; indeed short-baseline image clusters can degrade reconstruction accuracy significantly. An example of a good sequence of viewpoints (one picture from each viewpoint) is given in Fig. 4.7. Stereoscopic aerial photography is normally collected with 50–60% overlap between adjacent images, but for SfM-MVS the overlap is perhaps best considered in terms of both coverage and angular change. Each surface needs to be imaged at least twice, from different positions, but ideally positions that are not polar opposites such that the SIFT algorithm cannot successfully match features. Moreels

(a)

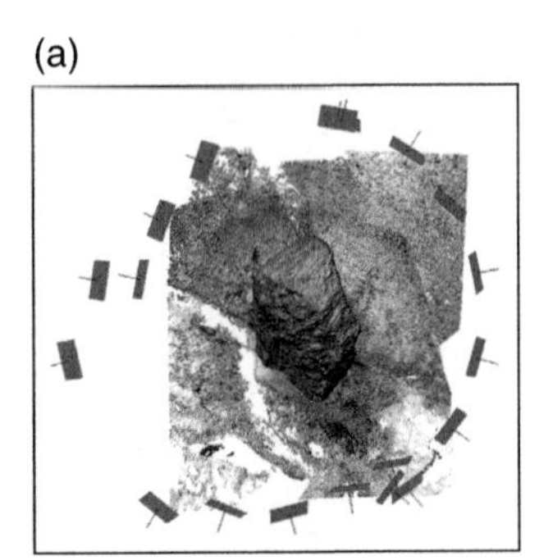

(b)

**Figure 4.7** Example of camera positions and orientations used in generating a 3D model of a glacially transported boulder: (a) top view and (b) side view.

and Perona (2007) suggest angular changes between images of greater than 25 and 30° should be avoided for exactly this reason. For objects of interest (i.e. that the user can circumnavigate) angular divergence between images should be limited to 10 and 20° if possible (Bemis et al. 2014).

In cases where the object of interest is planar (e.g. an exposed cliff section), the user should move along the feature taking adjacent images under near-parallel (and orthogonal) viewing conditions. However, the scale of the surface and the surrounding terrain may preclude this simple approach, and recent research has shown that systematic errors can also be introduced if all photos have parallel viewing directions (James & Robson 2014), so the inclusion of some imagery that is inclined relative to the viewing angle of the larger data set is advisable.

It is possible to process an SfM-MVS point cloud without applying any scale or position information (this would be a "relative" point cloud), but to extract useful information on size, distance, and volume it is necessary to acquire additional ground control. Scale can be added to a model very simply if the distance between two points is known. Greater accuracy and full 3D referencing require three or more GCPs, which are usually acquired by total station or differential GPS. It is a normal practice to ensure these points are acquired on stable features, such that they can be re-visited in future surveys for an accurate comparison between time-separated data. They should also be distributed widely across the area of interest, not neglecting the margins, and where possible avoiding any linear configuration. Various options exist for marking out GCP locations – in our experience consumer-grade survey targets, colour-coded bin sacks, small survey discs glued to larger plastic dishes, and sports cones can all make appropriate ground markers (Fig. 4.8). For close-range photogrammetry soluble fluorescent spray paint can also provide a temporary target. It should be noted, however, that the size and distinctiveness of the targets (with respect to their surroundings) can be a major limit on georeferencing accuracy. The case study in Box 4.1 evaluates the effects of the *distribution* and *density* of GCPs on SfM-MVS point cloud registration errors and accuracy and considers two questions: "What is the role of the ground control on data accuracy?"

**Figure 4.8** Examples of different target types used as ground control points.

An alternative method for using ground control data within the image is to have some method of georeferencing the camera locations for the input images. Presently, however, GPS units within consumer-grade hand-held cameras are of insufficient quality to provide the good-quality positioning data that would be required for fine-resolution structural mapping, for example, (Bemis et al. 2014) or other small-scale projects that are conducted over a few hundred metres or less.

## 4.5 Software

SfM-MVS software has developed rapidly in the past 5 years in line with advances in the field of computer vision. The underlying processing methods are well described in Chapter 3, so here we shall focus less on what is going on behind the screen and more on what each of these software can offer. Broadly speaking, SfM-MVS software fits into one of the following three categories (Torres et al. 2012):

**Box 4.1** Case study: Multi-temporal reach-scale topographic models in a wandering river – uncertainties and opportunities

Damià Vericat, Efrén Muñoz-Narciso, María Béjar, and Ester Ramos-Madrona
Fluvial Dynamics Research Group, Department of Environment and Soil Sciences, University of Lleida

### Background and context

Structure from Motion–Multi-View Stereo (SfM-MVS) is offering a new set of opportunities and challenges to geomorphologists. However, a question arises that deserves critical consideration: what is the role of ground control on data accuracy? Here we first evaluate the effects of the distribution and density of ground control points (GCPs) on the registration errors and accuracy of SfM-MVS point clouds by taking a terrestrial laser Ssanning (TLS) survey as a reference.

### Method

Multi-temporal SfM-MVS-based point clouds of a 12 km wandering river reach of the Upper Cinca (Southern Pyrenees) are obtained in the background of the research project MorphSed. A video fly-through of one of the point clouds is available at www.morphsed.es.

**Field data acquisition**. (a) SfM-MVS: Aerial photographs are taken at 275 m above the ground by means of a DSLR camera manually operated from a gyrocopter. Four flight paths are defined in an order to fully cover the 12 km long and 500 m wide river reach. Image resolution is around 0.05 m. (b) GCPs: A total of 220 GCPs are deployed regularly and RTK-GPS surveyed (total errors <0.05 m) before the flight is conducted. (c) TLS: A Leica C10 TLS provided a reference point cloud. A bridge (i.e. flat surface) and adjacent bedrock section (i.e. rough) were surveyed from two stations to minimise shadows. Registration errors were less than 0.02 m.

**Evaluation**. (a) SfM-MVS point clouds were obtained by means of Agisoft PhotoScan Professional; (b) the number of GCPs were reduced to evaluate the effects of their density and spatial distribution; (c) four spatial distribution patterns were established: perimeter, cross-section, grouped, and diagonal; (d) the extracted GCPs are used as check points to evaluate data accuracy; (e) TLS point clouds were registered using Cyclone 8, and the point clouds of flat and rough areas were exported; and (f) data accuracy was evaluated comparing SfM-MVS and TLS-based point clouds of flat and rough surfaces by means of Cloud Compare. SfM-MVS point clouds were filtered using ToPCAT to extract different topographic statistics at multiple grid sizes and converted to raster in ArcMap 10.1.

### Main findings

Results are summarised as follows:

1. The registration errors (expressed as root-mean-square error or RMSE) of the SfM-MVS-based point clouds do not change dramatically when GCPs are removed. The total error ranges between 0.15 and 0.20 m (Fig. B4.1ia).
2. Registration errors can be improved by removing suspicious GCPs. For instance, the total error changed from 0.20 to 0.14 m by removing 7 GCPs from the full set.

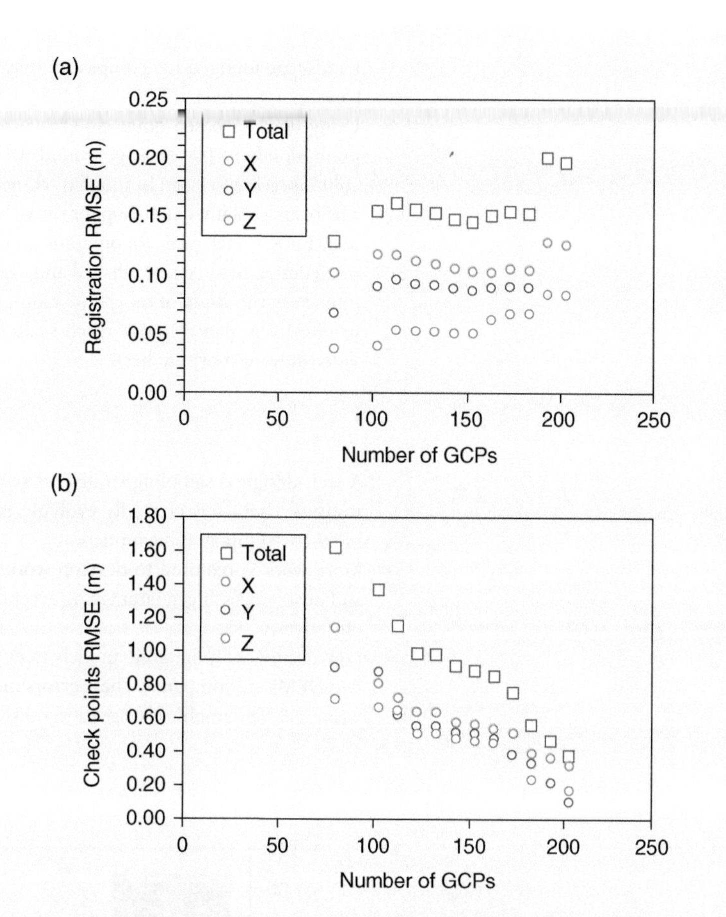


**Figure B4.1i** Registration errors associated with different GCP densities (a), and accuracy of point clouds as GCP density was reduced (b). Errors are expressed as the RMSE. Note that the accuracy is based on check points, and these are the GCPs that were excluded for the registration. For instance, the point cloud obtained by means of 100 GCPs had 120 check points. The RMSE of the GCPs provides the registration error, whilst the RMSE of the check points provided the accuracy of the point cloud. The reduction of GCPs was performed randomly at around 10% increments.

3 Although the registration errors remain similar, the analysis of the check points shows as the accuracy of the models clearly decreases as the number of GCPs is reduced (Fig. B4.1ib).
4 The RMSE of the check points ranges between 0.4 and 1.6 m. The minimum error is obtained when the majority of GCPs are used.
5 A set of check points is necessary to validate SfM-MVS-based point clouds even when low registration errors are obtained.
6 Analysis of the distribution of GCPs shows that restricting the location of these to around the perimeter of the area of interest maintains accuracies similar to those obtained with the whole set of GCPs and can thus maximise the survey efficiency.

7 Landscape topographic complexity may control the accuracy of SfM-MVS-based point clouds.
8 Using the TLS-based point cloud as a reference, the accuracy of SfM-MVS point clouds in flat areas is around 0.4 m (3D error), although this RMSE could increase to 0.6 m in rough or complex areas (Fig. B4.1ii).
9 The results obtained in the upper river Cinca show that the required field data to elaborate reach-scale topographic models can be obtained quickly if a sampling design has been established and evaluated (Fig. B4.1iii). Such data are of interest in the study of morphodynamics in fluvial systems and are essential to effectively parameterise reach-scale hydraulic models (Fig. B4.1iiic and video at www.morphsed.es).

### Key points for discussion

- A well-designed sampling strategy is required to maximise data accuracy to study or characterise rapidly evolving landscapes by means of repeat SfM-MVS-based topographic models.
- More work is required to develop workflows that optimise data acquisition and post-processing, guaranteeing acceptable accuracies in relation to specific objectives.
- Low-accuracy topographic models may yield a large propagated error when two DEMs are compared. These errors make very difficult the distinction between real topographic changes and noise.

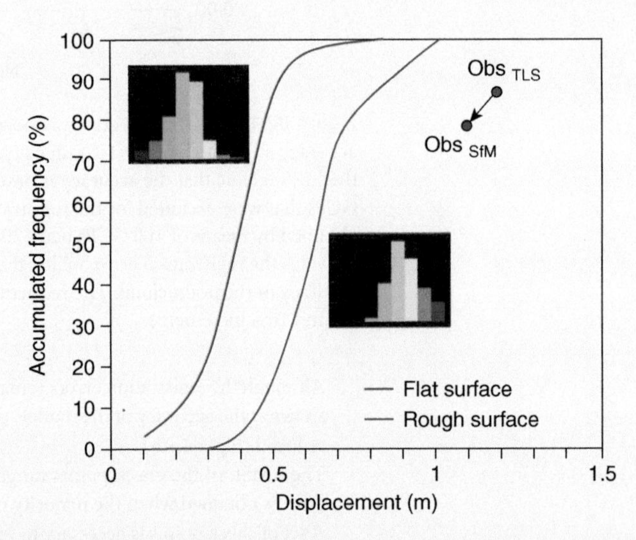


**Figure B4.1ii** The accuracy of SfM-MVS-based point clouds for rough and flat surfaces. The TLS point cloud of these surfaces was used as a reference or validation data set. The displacement vector between SfM-MVS and TLS observations was calculated in CloudCompare. Both histograms are presented as an inset.

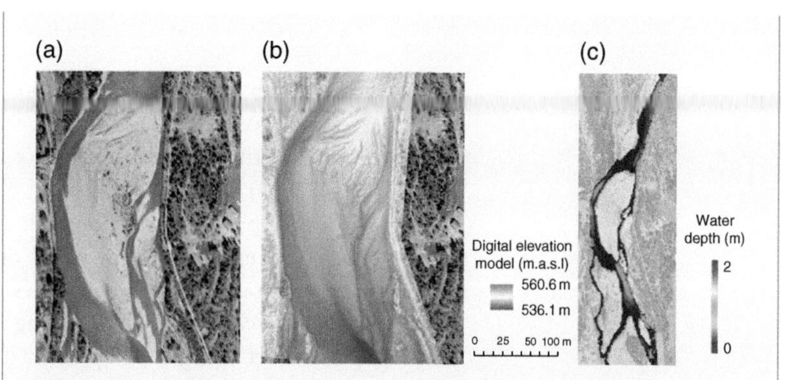


**Figure B4.1iii** (a) Orthophotograph of a central bar in the upper river Cinca and (b) SfM-MVS-based digital elevation model (DEM) of the same bar presented in (a). The point cloud was filtered by topographic point cloud analysis toolkit (ToPCAT) at a grid resolution of 0.5 m. (c) Hydraulic modelling using topographic products elaborated from the SfM-MVS point cloud of the upper river Cinca. Low flow conditions were modelled by Iber. The central bar in (c) is the same as in (a and b).

- SfM-MVS provides a way to extract high-density topography to support hydraulic modelling and characterise landforms at unprecedented spatial and temporal scales. The effect of relatively low data accuracy on the final results might be considered irrelevant in these cases.

### Summary

New advances in topographic data collection have substantially reduced the time involved in data collection and increased their spatial coverage. This new paradigm is essential for earth science studies. SfM-MVS poses a trilogy that opens new opportunities to geoscientists: speed/cost/resolution. However, data accuracy needs to be critically reviewed for the benefit of these opportunities to be maximised.

### Acknowledgement

MorphSed is funded by the Spanish Ministry of Economy and Competitiveness (CGL2012-36394) and the European Union (FEDER funds).

1. Individual algorithms that solve specific tasks within the reconstruction process;
2. Stand-alone tools, which solve the whole reconstruction process generating either a 3D mesh or a dense point cloud;
3. Web services to which images can be uploaded returning to the user the 3D model.

Available software can be further broken down into those that are commercially developed and that are open-source.

### 4.5.1 Commercial Software

**123D Catch/ReCap** (http://www.123dapp.com/catch) is a freely available cloud-based application that has found great use in cultural heritage documentation, architecture, and archaeology (e.g. Rasztovits & Dorninger 2013; Santagati et al. 2013). It uses either a stand-alone installation on the user PC or a smartphone app to manage photo uploads, manual improvement of the initial point cloud extraction, download of the improved point cloud or mesh, and export to a variety of data types: *.dwg, *.fbx, *.rzi, *.obj, *.ipm, *.las. It also has the capability of generating video footage of the feature and printing the project using a 3D printer. The software has limited post-processing tools, meaning that noise filtering or hole filling must be done externally, it can only handle images taken from a single sensor within a project, and its processing time is between 1 and 2 hours based on an input data set of approximately 70 images (Butnariu et al. 2013). In the geosciences it has been used to estimate gulley headcut erosion (Gómez-Gutiérrez et al. 2014a), quantify changes in rock glacier dynamics (Gómez-Gutiérrez et al. 2014b), and characterise alpine fluvial and hillslope geomorphology (Micheletti et al. 2014) and coastal geomorphology (James et al. 2013).

**AgiSoft PhotoScan** (http://www.agisoft.com/) is one of the major commercial SfM-MVS softwares currently available. It is a stand-alone application that is available at a cost of GBP2300 for the professional edition, or GBP118 for the standard edition. Agisoft provides educational licenses at a cost of GBP363 (professional) or GBP40 (standard). The standard edition allows users to triangulate their photos, generate, and edit a dense point cloud and generate and texture a 3D model. Professional users have the added capability of classifying the point cloud, exporting DEMs, creating orthophotograph mosaics, scripting process chains, and processing multi-spectral imagery. Post-processing tools are again limited to basic editing operations, but the software exports to six widely used formats (*.obj, *.txt, *.las, *.ply, *.u3d, and *.pdf) so that other code can be used to filter and fill derived point cloud data. Exports from PhotoScan are also possible to (i) interactive pdf, where zoom, pan, and rotate functions can be performed on the model and to (ii) "sketchfab" (via "upload model" after a texture has been built) for sharing of interactive files and videos. PhotoScan has major advantages over competing software in terms of user friendliness, but can be quite computationally expensive; for large projects (incorporating >500 images), it is recommended to employ a 64-bit operating system with at least 6 Gb of RAM (Neitzel & Klonowski 2011). Even with this computational power, processing imagery on the high-accuracy settings can routinely take of the order of 24 hours. PhotoScan has been applied to a wide range of geoscience projects, for example, measuring coral reef terrain roughness (Leon et al. 2015) and modelling braided river topography (Javernick et al. 2014), and it is also extremely popular for archaeology, cultural heritage, and many other disciplines and hobbies, as evidenced by the many examples online at "sketchfab" (search "Agisoft" or "PhotoScan").

**Pix4DMapper** (http://pix4d.com/) is an emerging software developed by a spin-off company from the École Polytechnique Fédérale de Lausanne. It is stand-alone and available at a one-off cost of GBP4810, or to rent for GBP1925 per year or GBP192 per month. One-off educational licenses are available for GBP1110 (single) or GBP3700 (25 devices). It has undergone limited testing within the scientific literature, but the evidence that is available suggests it performs well in bundle adjustment but less well in dense matching and orthophotograph generation (Unger et al. 2014). It has been successfully employed to measure rill and inter-rill erosion on loess soils (Eltner et al. 2014) and for automated gully mapping (Castillo et al. 2014), but otherwise its use in geoscientific applications has been limited.

**Autodesk ImageModeler** (www.autodesk.com/imagemodeler) was originally offered as a stand-alone product but is now only available with the purchase of a license for other Autodesk products. It is partly marketed on its ease of use, with a three-step workflow – calibration, modelling, and texturing – comprising the core of the software. Its primary application is in architectural visualisation, and as such it has seen little use in the geosciences, perhaps because the camera interior orientation is accomplished by manual feature matching of common points between each pair of adjacent pictures. This can be a laborious step that some competitor software does not require.

**D-Sculptor** (http://www.d-vw.com/) also requires manual feature matching in the early processing stages. The software was originally developed for the modelling of discrete objects, such as ceramics and pottery, and a major processing step requires the user to mask around the object of interest in every photograph, which can become unfeasible for large data sets. It is available for around GBP500, but given its dependence on manual interaction it has found little application in the geosciences.

**PhotoModeler** (http://www.photomodeler.co.uk/) is another software that depends on some initial manual feature matching. It is developed and distributed by Eos Systems and has seen application in underwater archaeological surveying, facial image identification, and accident scene reconstruction. As with these examples, its use in the geosciences has been limited by the requirement for manual matching, making it unfeasible for large data sets. Applications have focussed on close-range photogrammetry where small data sets are common and include measuring gully morphological evolution (Gesch et al. 2015) and measuring plot-scale glacier surface roughness (Irvine-Fynn et al. 2014).

**Microsoft Photosynth** (https://photosynth.net/) is a Web-based implementation of Bundler (see Section 4.5.2), performing the SIFT analysis remotely and posting the SfM-MVS point cloud online for the user to view, interrogate, and, if required, download. It was first released in 2006, following several years of collaboration between Microsoft and the University of Washington. In 2008, it was officially released to the public, and within 1 year it had processed over 400,000 synths (clouds). Since 2012, Photosynth has also been available as an app for smartphones, which can either be used to generate panoramas, or as with the Web-based version, 3D point cloud data. The resulting point cloud is sparsely populated so dense clouds must be extracted using more advanced software such as PMVS2 (see later text). Similarly, Photosynth has limited processing/editing/cleaning tools, so most subsequent processing of Photosynth clouds usually takes place in software such as Meshlab (http://meshlab.sourceforge.net/) – an open-source application for processing and editing 3D triangular meshes. A further limitation of Photosynth is that camera positions and attitude are calculated in a relative coordinate system, meaning absolute positioning is required *a posteriori*. The use of Photosynth in the geosciences remains fairly limited to date. Examples include deriving river reach topography (Fonstad et al. 2013) and identifying structural controls in an active lava dome (James & Varley 2012).

**ARC3D** (Automatic Reconstruction Cloud; http://www.arc3d.be/) provides a Web service to which users can upload their images and then return to the site at a later point to download the generated point cloud. The user is provided with dense depth maps for every image and the corresponding camera parameters. As with point cloud data produced using Microsoft Photosynth (earlier), to be able to register image depth maps into complete surfaces, and then mesh, refine, and simplify them, the user requires Meshlab (http://meshlab.sourceforge.net/). Both of these softwares (ARC3D and Meshlab) have been widely used in cultural heritage projects.

### *4.5.2 Open Source Code (Academic Development)*

**Bundler** (http://www.cs.cornell.edu/~snavely/bundler/) was developed at the University of Washington by Noah Snavely, and it is a freely available stand-alone application. It was first developed to take advantage of large online image collections to construct photo tours of scenic or historic locations in 3D (Snavely et al. 2006). Critically, Bundler produces sparse point clouds, with the intention that dense cloud data can be derived through related open-source software such as PMVS and CMVS (see later text). As well as the derived point cloud data, users are also provided with a *.ply file that contains reconstructed cameras and points. The main disadvantage of the Bundler software is that it does not exploit the graphic processing units available on contemporary desktop PCs, using instead only a single-processor core at any one time. Consequently, processing time can be slow in comparison to more recent software (i.e. Agisoft PhotoScan; Turner et al. 2014). Studies have shown that whilst Bundler performs very well (in terms of spatial accuracy) in photo-tourism applications, it is less suited to applications using a single camera with a fixed focal length (Rosnell & Honkavaara 2012), because it assumes that each image is taken from a different camera and re-calculates camera calibration parameters with each image accordingly. It should therefore be used with caution in aerial mapping surveys, for example.

**Patch-based MVS (PMVS)** (http://www.di.ens.fr/pmvs/) is MVS software only, so it is usually used in conjunction with other SfM-MVS-based software (e.g. Bundler) capable of estimating the interior and exterior orientation of the scene. It was originally developed by Yasutaka Furukawa at the University of Washington, and it is now on its second version (PMVS2). PMSV2 outputs oriented points (rather than a polygonal (or a mesh) model), where both the 3D coordinate and the surface normal are estimated at each oriented point. PMVS2 is currently packaged with clustering views for MVS (CMVS), which attempts to overcome the computational and memory resources required to process large data sets (see Section 3.9). It does this by decomposing input images into sets of clusters that are processed either individually or in parallel. It also removes redundant images from the processing chain, leading to a faster and more accurate reconstruction. PMVS/PMVS2 are widely used in the academic literature; soil erosion monitoring (Aucelli et al. 2012), mineral extraction monitoring (Wang et al. 2014), and coastal geomorphology (James et al. 2013) are some of the geoscience disciplines in which PMVS has been used.

**VisualSFM** (http://ccwu.me/vsfm/), developed by Changchang Wu at the University of Washington, is a spin-off from Bundler, and it incorporates PMVS/CMVS into its graphical user interface (GUI) so that dense reconstructions can be derived as part of a single workflow (Wu 2013). It is computationally fast (being multi-core accelerated) and has found application in cliff-face reconstruction (Dewez 2014) and landslide monitoring (Stumpf et al. 2015).

**Apero MicMac** (http://logiciels.ign.fr/?-Micmac,3-) was developed by the French public state establishment – Institut National de l'Information Géographique et Forestière (IGN). It comprises three modules: (i) initial feature extraction using SIFT, (ii) automatic computation of image orientation, and (iii) dense point cloud extraction. It is one of the more complex softwares to use but has been shown to offer a more accurate and more complete result than some other software (Bretar et al. 2013) because of its rigorous camera calibration model (Ouédraogo et al. 2014). It has been widely applied to architectural studies (e.g. Deseilligny & Clery 2011) and more recently to geoscience applications such as measuring the surface roughness of volcanic terrains (Bretar et al. 2013) and creating forest canopy height models (Lisein et al. 2013).

## 4.6 Point Cloud Viewers

**CloudCompare** (http://www.danielgm.net/cc/) is a popular, open-source, point cloud processing software widely used amongst academics having been originally developed as part of a joint industrial PhD project between Telecom Paris and EDF Energy. Its original application was in the comparison of point clouds acquired by laser scanning, mostly for structural analyses of industrial facilities (i.e. power plants) or building sites. It comprises a suite of tools for editing (cleaning) and rendering point cloud and triangular mesh data, as well as more advanced algorithms for projection transformations, registration, distance computation, statistical analysis, segmentation, and the estimation of various geometric parameters (e.g. density and roughness).

**PolyWorks IMView** (http://www.innovmetric.com/en/products/polyworks-viewer) was primarily designed to render 3D data derived from the IMInspect module of the commercial version of PolyWorks software. It is, however, freely available, and is capable of handling point clouds, meshes, and CAD data set. It contains basic analysis tools to measure distances, radii, angles, and cross-sections if the data originates from PolyWorks software, but if .obj, .wrl, or .ply data are imported, the measurement functionality is disabled and the software acts as a viewer only.

**Geomagic Verify Viewer** (formerly Rapidform Explorer; http://www.rapidform.com/products/xov/explorer-free-viewer/) is another freely available data viewer with measurement (distance, angles, radii, area, and volume) capability. It takes files generated by its sister software, Verify, as well as other popular 3D formats such as .stl, .obj, .igs, and .stp. It is also easy to embed highly compressed 3D models into Web pages using this viewer.

**Meshlab** (http://meshlab.sourceforge.net/) is an open-source software for viewing, processing, and editing point cloud and 3D triangular meshes. It was developed at the Visual Computing Lab in Pisa, and it is often used in a workflow that comprises Bundler and PMVS2 in preceding steps. Functions include mesh creation, editing, cleaning, healing, and inspection, and it can handle a multitude of import and export file types. It has basic measurement capabilities, allowing linear measurements, as well as more advanced tools for curvature analysis and registration.

Many more freely available point cloud viewers exist and, although not comprehensive, the most popular are summarised in Table 4.4.

It is also worth noting that most commercial softwares distributed for laser scanners (e.g. Leica Cyclone and Riegl RiScan Pro) provide another option for viewing SfM-MVS-derived point cloud data, most of which have the added capability of registration, editing, and filtering functionality.

## 4.7 Filtering

SfM-MVS data rarely have homogeneous point densities, and small errors in matching can lead to numerous outliers that can corrupt subsequent analyses. For example, a noisy data set (i.e. one with many outliers) can produce spurious first-order calculations, such as surface normals or curvature changes. Sometimes there are clear (and often discrete) blunders that require removal. Sometimes there are holes in the data that require filling. And sometimes the cloud is so dense that it simply requires the systematic

**Table 4.4** Freely available software for viewing point-cloud data.

| Viewer name | URL | Notes |
|---|---|---|
| Bentley Pointools View | http://bit.ly/1Sn4Wx0 | Basic measuring and editing functionality, for example, clipping, sketching, and annotations |
| FugroView | http://www.fugroviewer.com/ | Can create contours, TINs, and cross-sectional data |
| ccViewer | http://www.danielgm.net/cc/ | Is a light version of CloudCompare, without the editing functionality |
| 3DReshaper Viewer | http://www.3dreshaper.com/en1/En_download-free-3d-viewer.htm | Basic measuring and labelling functionality (e.g. with image intensity); automatic point cloud reduction capability |
| LAStools lasview | http://www.cs.unc.edu/~isenburg/lastools/ | Has some editing functionality and can produce TINs and cross-sectional data |
| Trimble RealWorks Viewer | http://www.trimble.com/3d-laser-scanning/realworks.aspx | Measuring and down-sampling functionality included. Clipping possible and can create cross-sectional data |
| Global Mapper | http://www.bluemarblegeo.com/products/global-mapper.php | Exceptional support for different formats. Many measurement tools (e.g. distance, area, and cut-and-fill) |
| LiMON Viewer | http://www.limon.eu/products/16-limon-viewer | Basic measurement tools such as distance and area. Can generate cross-sectional data |
| LP360 Viewer | http://qcoherent.com/evaluation.html | Is an extension for ArcGIS and intended for analysing laser scanner data. Advanced features include automatic ground classification and building footprint extraction |

removal of redundant data so that processing times are more manageable. In practice, therefore, raw point clouds almost always require some form of filtering. For any of these cases a range of filters exist, many of which were originally designed for use with laser scanner data but can also be validly applied to SfM-MVS-derived point clouds.

One of the fastest-growing open-source libraries of filters is available from www.pointclouds.org. The Point Cloud Library (PCL) was initiated in early 2011, and has since been released (free of charge) for both commercial and research use. As well as providing filters, the library can offer surface reconstruction, registration, segmentation, and model fitting. The following description of filters in this section describes the basic functionality of these algorithms, based partly on a wiki maintained by the Robotics Group at the University of Léon (http://robotica.unileon.es/).

A **pass-through** filter is a useful method for rapidly removing background noise from a scene, for example, surrounding hillslopes when the point of interest is a river reach and its immediate banks. It takes a user-specified range in a given dimension and keeps only the points that fall within that range. Problems arise, however, when the reference frame of the cloud is oblique to the points that require removal. For example, filtering on the $Y$ value to remove all points above or below a planar surface (e.g. a floodplain) may yield undesirable results if the reference frame is at a non-orthogonal angle.

A similar tool (or at least one that can produce similar results) is a **conditional** removal filter. This requires some test to be performed on each point, and the point is either kept or removed based on whether the condition is met. For example, a user may set up a condition that points have to be at a certain elevation to be considered as part of a tree canopy; all points not meeting this criterion would then be removed. Users can build multiple conditions into queries. Continuing the tree canopy theme, users may require points to be at a certain elevation, AND within a known area of conservation, OR close to a nesting site, to be maintained. Boolean algebra is the key component here, and users can exploit other common operands such as LT, GT, EQ, GE, and LE, for example.

Outliers are points that have been incorrectly measured and are commonly referred to as noise. They lie somewhere distant to the main cloud so are relatively easily identified, and they can introduce significant error into subsequent calculations if not removed from the data set. Trimming them can reduce processing times significantly, not only because there are fewer points in the data set but also because the geometric space over which computation subsequently takes place is much smaller.

A **radius-based** outlier removal (Fig. 4.9) represents the simplest approach. Each point is queried, and the number of points within a user-specified distance determines whether it is labelled as an outlier (or not). Clearly, true outliers are lonely points, and thus have a few (if any) neighbours. Those in the main point cloud will return many neighbours, and thus will not be filtered. Although very effective, this filter can be computationally expensive when applied to large clouds. A **statistically based** alternative for removing outliers is to look at the distance of each point to its neighbour (or neighbours) and remove all points outside of the global mean plus one standard deviation. This depends on the data approaching a Gaussian distribution, which is a safe assumption in the majority of cases.

In some cases the point cloud is simply too dense to be processed by the available computing power. In this case, the user may require a **downsampling**

**Figure 4.9** Before (a) and after (b) applying a radius-based outlier removal filter. Because outliers tend to be lonely relative to non-outlier points, their removal based on a neighbourhood analysis is relatively easy.

filter. Downsampling filters work by gridding the available data in three dimensions (where each gridded cube is known as a voxel), and the points within each voxel are reduced to a single point based usually on the mean, max, median, min, or some weighted average based on the distance of each point from the voxel centroid. The voxel dimensions can be set by the user based on the intended application, and in most cases the user may stipulate there needs to be a minimum number of points within each voxel, otherwise it is assigned a non-applicable number (NaN). This process is commonly known as decimation, and we will look at it again when we consider DEM construction later. Conversely, an **upsampling** filter will seek to interpolate new voxel values where there are missing data, normally by some method of interpolation based on values in immediately surrounding voxels or by fitting a 3D surface to the data over much longer path lengths.

One of the most common uses of filtering software is to decimate point clouds to remove redundant data. A clear area for future research to focus on is the development of more **intelligent** decimation algorithms that take into account the existing homogeneity/heterogeneity of the surface that the filter is applied to. For example, file sizes could be significantly reduced if areas of topographic homogeneity were represented by fewer points – in geoscience this may relate to a bedrock surface, for example. In contrast, areas of topographic heterogeneity (edges for example) would not be filtered at all using this approach. There is also scope to filter point-cloud data depending on the colour of the feature within the images. Here, an obvious application may be to filter out water surfaces when constructing SfM-MVS data of river reaches. Because the water surface is a dynamic feature, point cloud data within the channel are usually spurious and are manually removed (and replaced with an interpolated surface) before flood routing models are applied, for example. Where large areas of undesirable data are included in point clouds, the user usually has the option of masking such features out before the dense cloud is extracted; using an intelligent filtering algorithm such as the one described here could expedite the process considerably. Such developments would closely follow on from point cloud classification methods, such as those presented in Brodu and Lague (2012), for example.

Whilst filtering methods are becoming increasingly intelligent, it should also be noted that **manual** filtering is always an alternative to these automated approaches. Particularly with small data sets, "cleaning" the point cloud by hand can offer an equally rigorous approach and gives the user full control over what is considered redundant or incorrect data; in many cases automated filtering algorithms can remove good and bad data in equal measure. There is no doubt that ensuring a point cloud data set is of sufficient quality to derive higher-level information is one of the most time-consuming tasks for SfM-MVS users; indeed it has been estimated that filtering and quality control combined can account for 60–80% of point-cloud processing time (Flood 2001).

## 4.8 Generating Digital Elevation Models from Point Clouds

Given the increasingly fine resolution of imaging sensors, and the ever-improving ability of SfM-MVS algorithms to match features from a variety of angles, it is inevitable that in some cases, the imaged surfaces are reconstructed at a resolution well beyond that required by the user. In these cases, the redundant data can either be culled in three dimensions (see the downsampling filter referred to earlier) or even decimated and transformed to two-dimensional (2D) data in a single step, thus creating a DEM, digital surface model (DSM), or digital terrain model (DTM). Although used interchangeably, these terms do differ in their meaning: DSMs generally refer to topographic data that include the surficial cover (e.g. trees and buildings); DTMs represent a "bare earth" surface, with surficial cover removed; DEMs are usually unspecified in terms of what they represent – and may thus show surficial, surface, or even sub-surface topography. Generally speaking, SfM-MVS data are initially gridded as DSMs, and the user then requires a 2D-filtering algorithm to remove the surficial cover if a DTM is required.

In the geosciences, there are several issues that the user needs to be aware of in making this transition from 3D to 2D data. The first is that the chosen grid size (or spatial resolution) of the derived DEM should be appropriate for the desired application. This can be a trade-off between keeping the resolution fine enough such that the terrain surface is adequately represented, but coarse enough such that subsequent analyses (e.g. flood routing) are not unfeasibly computationally expensive. Related to this (and the second issue to be aware of) is that when deriving geomorphometric information from DEM data, it should be noted that for every order of differentiation, the effective resolution of the product declines and noise increases, that is, small errors in the primary data set become increasingly exaggerated (Quincey et al. 2014). The third issue is that outliers and/or matching blunders can significantly impact elevation values, particularly when the number of points within a given cell are small and are almost impossible to detect once they have been decimated. Though it may seem an obvious point, it is thus critical that point-cloud data have been cleaned and, if necessary, filtered, prior to their decimation. Fourth, and finally, in transforming 3D data to a DSM, users may find that useful information is lost, normally in the *z*-axis. For example, there may be several surfaces represented within the same x, y grid cell (Fig. 4.10), meaning the user has a difficult choice to make as to which surface is represented in the DSM.

This last point, that data can be lost in reducing 3D point cloud data to two dimensions, can be offset by the opportunity to explore the data in more depth. In making the transformation, some softwares will calculate meaningful statistics for each grid cell. For example, the topographic point cloud analysis toolkit (ToPCAT; Brasington et al. 2012; Rychkov et al. 2013), which is available at http://gcd.joewheaton.org/, allows users to calculate zMin for modelling bare earth elevations, zMax for modelling surficial topography,

**Figure 4.10** An overhanging bedrock surface that when gridded in planform could have any one of the three surfaces highlighted as being the elevation of the associated grid cell.

and novelly detrended standard deviation, which can act as a proxy for surface roughness (Brasington et al. 2012). As with voxel-based methods, the user can specify a minimum number of points within each cell for it to be assigned an elevation value, meaning individual (often spurious) points do not skew any subsequent analysis, and users can easily take the output .txt file and grid the data in any GIS software. The attraction of this approach is that the user has full control over the transformation process, in contrast to many other algorithms (e.g. those in ArcGIS) that perform a single transformation function without the user necessarily having full knowledge of how the data are being reduced.

## 4.9 Key Issues

This chapter has not touched on data accuracy and how it is assessed, because error quantification is the main focus of the following chapter. Nevertheless, it should be emphasized that the foundations for making a rigorous assessment of data quality are laid during practical data capture as described here. For example, the collection of check points (i.e. additional GCPs that are not used in the point cloud scaling and registration process) with which the accuracy of the cloud can be independently assessed, is a necessary but often neglected part of the data collection process. It should also be emphasized that the quality of the input data (images and ground control) will be a major determinant of the error that propagates through to the final 3D model. Therefore data capture and error quantification (see Chapter 5) should be viewed as intimately connected parts of the SfM-MVS workflow rather than discrete stages.

If topographic data derived from SfM-MVS are to be truly integrated into geoscience applications (e.g. surface process modelling), and results compared between independent studies, there is a requirement for a representational framework that scientists can work towards. Currently, scientists employ a range of analytical tools, algorithms, processing approaches, and software for the generation, manipulation, and interpretation of these topographic data. Methods are highly empirical, thus the type and quality of the derived data are dependent, to a large extent, on the analyst. Consequently, replication of existing results can be difficult. Standardisation and protocols for information extraction and integration are therefore required if data quality, result accuracy, and the validity of comparing measurements across different studies (and study areas) are to be assured.

## 4.10 Summary

This chapter has presented an overview of the platforms, sensors, and software that are available for SfM-MVS users, as well as providing basic advice regarding image and ground control capture to ensure the resulting point cloud data are as accurate as possible. Platforms used to mount a camera include mast, pole, or boom, UAVs, kites, lighter-than-air balloons, and manned aircraft. Sensors include stills cameras, DSLRs, those in mobile phones, video cameras, and trail cams. SfM-MVS software fits into one of the following three categories:

1. Individual algorithms that solve specific tasks within the reconstruction process;
2. Standalone tools, which solve the whole reconstruction process generating either a 3D mesh or a dense point cloud;
3. Web services to which images can be uploaded returning to the user the 3D model.

Software for viewing and processing point clouds will be useful to geoscientists applying SfM-MVS and wishing to manually edit, automatically filter, or decimate their data, as well as to convert it to a DEM.

The possibility for enthusiasts to get involved in processing topographic data at zero cost (providing a basic camera is available) is one of the most attractive features of SfM-MVS. SfM-MVS simply requires multiple images taken from different positions of a target object or surface of interest. However, this apparent ease of acquisition of images must be tempered with note of some subtle yet crucial considerations; indeed these considerations might be termed as guidelines for best practice. For example, it is not sufficient to have a lot of pictures from a single location.

For close-range SfM-MVS, pictures taken at each (human) step whilst moving around the object of interest are ideal, but these images must cover the top, side, and underside (if present) of the object. For far-range SfM-MVS such as when deriving models of the physical environment, images should be acquired from positions situated around the landscape "looking inwards" to provide 360° coverage. For example, images taken around a horseshoe ridge line of a catchment can provide excellent data. Inevitably, ground-based camera positions are likely to be restricted to several high-elevation but well-spaced viewpoints. At large scales, and when working in the physical environment, recent work has shown that the following three main parameters affect the quality of the 3D model:

1. Lighting conditions;
2. Changes in shadow length and surface albedo;
3. Number and spatial arrangement/distribution of camera positions.

Planning a field survey for SfM-MVS is therefore critical. Where possible, images for SfM-MVS should be

- Acquired under constant lighting;
- Where moving shadows and camera flash are minimised;
- Where the object of interest is fixed;
- Where the object or surface of interest contains variations in texture and colour;
- Where the target object or surface is imaged with overlapping images taken at a range of angles from that object or surface.

Some of these issues can be overcome by the use of a camera with a short focal length to reduce the number of images (and thus time) required, but there is an associated reduction in the resolution of the generated model.

Theoretically, SfM-MVS can work with only three input images, but the density of a point cloud tends to be positively correlated with the number of good-quality images used. Practically, more than a few hundred images in a single batch will create problems with computing power. Additionally, it is useful to down-sample images for computational efficiency; high-resolution images are only necessary for images with a long base line to the target surface; 3 MP is apparently ample for close-range SfM-MVS.

With SfM-MVS ideal image overlap must be considered in terms of both coverage and angular change. Each surface needs to be imaged at least twice, from different positions. A consensus is that angular divergence between overlapping images should be approximately 10–20° if possible. In cases where the object or surface of interest is planar and images are (perhaps necessarily) taken virtually entirely perpendicularly, systematic errors can be introduced (James & Robson 2014) so the inclusion of some imagery that is inclined relative to the viewing angle of the larger data set is advisable.

Overall, whilst many of the SfM-MVS software available are black box and thus preclude rigorous assessments of processing parameters and their impact on point cloud accuracy, there is a growing focus in the academic community on validating SfM-MVS-derived data with some form of ground truth, whether from traditional point-based survey methods (e.g. dGPS) or alternative point cloud data (e.g. from laser scanning). Chapter 5 is devoted to this challenge.

## References


Aucelli, P.P.C., Conforti, M., Della Seta, M. et al. (2012) Quantitative assessment of soil erosion rates: results from direct monitoring and digital photogrammetric analysis on the Landola catchment in the Upper Orcia Valley (Tuscany, Italy). *Rendiconti Online Società Geologica Italiana*, **21**, 1199–1201.

Bell, A.D.F. (2012) Section 2.3.1: Creating DEMs from survey data (interpolation methods and determination of accuracy). In: L.E. Clarke & J.M. Nield (eds), *Geomorphological Techniques (Online Edition)*. British Society for Geomorphology, London.

Bemis, S.P., Micklethwaite, S., Turner, D. et al. (2014) Ground-based and UAV-based photogrammetry: a multi-scale, high-resolution mapping tool for structural geology and paleoseismology. *Journal of Structural Geology*, **69**, 163–178.

Bendig, J., Bolten, A. & Bareth, G. (2012) Introducing a low-cost mini-UAV for thermal and multispectral imaging. *International Archives of the Photogrammetry, Remote Sensing and Spatial Information Sciences*, **XXXIX-B1**, 345–349.

Boike, J. & Yoshikawa, K. (2003) Mapping of periglacial geomorphology using kite/balloon aerial photography. *Permafrost and Periglacial Processes*, **14**, 81–85.

Brasington, J., Vericat, D. & Rychkov, I. (2012) Modeling river bed morphology, roughness, and surface sedimentology using high resolution terrestrial laser scanning. *Water Resources Research*, **48 (11)**, W11519.

Bretar, F., Arab-Sedze, M., Champion, J., Pierrot-Deseilligny, M., Heggy, E. & Jacquemoud, S. (2013) An advanced photogrammetric method to measure surface roughness: application to volcanic terrains in the Piton de la Fournaise, Reunion Island. *Remote Sensing of Environment*, **135**, 1–11.

Brodu, N. & Lague, D. (2012) 3D terrestrial lidar data classification of complex natural scenes using a multi-scale dimensionality criterion: applications in geomorphology. *ISPRS Journal of Photogrammetry and Remote Sensing*, **68**, 121–134.

Butnariu, S., Gîrbacia, F. & Orman, A. (2013) Methodology for 3D reconstruction of objects for teaching virtual restoration. *International Journal of Computer Science Research and Application*, **3 (1)**, 16–21.

Castillo, C., Taguas, E.V., Zarco-Tejada, P., James, M.R. & Gómez, J.A. (2014) The normalized topographic method: an automated procedure for gully mapping using GIS. *Earth Surface Processes and Landforms*, **39 (15)**, 2002–2015.

Chao, H., Coopmans, C., Di, L. & Chen, Y.Q. (2010) A comparative evaluation of low-cost IMUs for Unmanned Autonomous Systems. In: *IEEE International Conference on Multisensor Fusion and Integration for Intelligent Systems*, September 5–7, 2010, pp. 211–216. University of Utah, Salt Lake City, UT.

Church, M., Hassan, M.A. & Wolcott, J.F. (1998) Stabilizing self-organized structures in gravel-bed stream channels: field and experimental observations. *Water Resources Research*, **34 (11)**, 3169–3179.

Colomina, I. & Molina, P. (2014) Unmanned aerial systems for photogrammetry and remote sensing: a review. *ISPRS Journal of Photogrammetry and Remote Sensing*, **92**, 79–97.

Deseilligny, M.P. & Clery, I. (2011) Apero, an open source bundle adjustment software for automatic calibration and orientation of set of images. *ISPRS-International Archives of the Photogrammetry, Remote Sensing and Spatial Information Sciences*, **38**, 5.

Dewez, T. (2014) Reconstructing 3D coastal cliffs from airborne oblique photographs without ground control points. *ISPRS Annals of the Photogrammetry, Remote Sensing and Spatial Information Sciences*, **2** (**5**), 1.

Dunford, R., Michel, K., Gagnage, M., Piégay, H. & Trémelo, M.-L. (2009) Potential and constraints of Unmanned Aerial Vehicle technology for the characterization of Mediterranean riparian forest. *International Journal of Remote Sensing*, **30** (**19**), 4915–4935.

Eisenbeiss, H. & Zhand, L. (2006) Comparison of DSMs generated from mini UAV imagery and terrestrial laserscanner in a cultural heritage application. *The International Archives of Photogrammetry, Remote Sensing and Spatial Information Sciences*, **XXXVI** (part 5), 90–95.

Eltner, A., Baumgart, P., Maas, H.-G. & Faust, D. (2014) Multi-temporal UAV data for automatic measurement of rill and interrill erosion on loess soil. *Earth Surface Processes and Landforms*. doi: 10.1002/esp.3673.

Flood, M. (2001) LiDAR activities and research priorities in the commercial sector. *International Archives of Photogrammetry Remote Sensing and Spatial Information Sciences*, **34** (3/W4), 3–8.

Fonstad, M., Dietrich, J., Courville, B., Jensen, J. & Carbonneau, P. (2013) Topographic Structure from Motion: a new development in photogrammetric measurement. *Earth Surface Processes and Landforms*, **38**, 421–430.

Gesch, K.R., Wells, R.R., Cruse, R.M., Momm, H.G. & Dabne, S.M. (2015) Quantifying uncertainty of measuring gully morphological evolution with close-range digital photogrammetry. *Journal of the Soil Science Society of America*, **79**, 650–659.

Gómez-Gutiérrez, A., Schnabel, S., Berenguer-Sempere, F., Lavado-Contador, F. & Rubio-Delgado, J. (2014a) Using 3D photo-reconstruction methods to estimate gully headcut erosion. *Catena*, **120**, 91–101.

Gómez-Gutiérrez, A., Sanjosé-Blasco, J.J., Matías-Bejarano, J. & Berenguer-Sempere, F. (2014b) Comparing two photo-reconstruction methods to produce high density point clouds and DEMs in the corral del veleta rock glacier (Sierra Nevada, Spain). *Remote Sensing*, **6** (**6**), 5407–5427.

Irvine-Fynn, T.D., Sanz-Ablanedo, E., Rutter, N., Smith, M.W. & Chandler, J.H. (2014) Measuring glacier surface roughness using plot-scale, close-range digital photogrammetry. *Journal of Glaciology*, **60** (**223**), 957–969.

James, M.R. & Robson, S. (2012) Straightforward reconstruction of 3D surfaces and topography with a camera: accuracy and geoscience application. *Journal of Geophysical Research*, **117**, F03017.

James, M.R. & Robson, S. (2014) Mitigating systematic error in topographic models derived from UAV and ground-based image networks. *Earth Surface Processes and Landforms*, **39** (**10**), 1413–1420.

James, M.R. & Varley, N. (2012) Identification of structural controls in an active lava dome with high resolution DEMs: Volcán de Colima, Mexico. *Geophysical Research Letters*, **39** (**22**). doi: 10.1029/2012GL054245.

James, M.R., Ilic, S. & Ruzic, I. (2013) Measuring 3D coastal change with a digital camera. In: *Proceedings of Coastal Dynamics 2013,* June 24–28, 2013, pp. 893–904. France.

Javernick, L., Brasington, J. & Caruso, B. (2014) Modeling the topography of shallow braided rivers using Structure-from-Motion photogrammetry. *Geomorphology*, **213**, 166–182.

Klein, G. & Murray, D. (2009) Parallel tracking and mapping on a camera phone. In: *Mixed and Augmented Reality, 2009*. ISMAR 2009, October 19–22, pp. 83–86. Orlando, FL.

Lejot, J., Delacourt, C., Piégay, H., Fournier, T., Trémélo, M.-L. & Allemand, P. (2007) Very high spatial resolution imagery for channel bathymetry and topography from an unmanned mapping controlled platform. *Earth Surface Processes and Landforms*, **32**, 1705–1725.

Leon, J.X., Roelfsema, C.M., Saunders, M.I. & Phinn, S.R. (2015) Measuring coral reef terrain roughness using "Structure-from-Motion"close-range photogrammetry. *Geomorphology*, **242**, 21–28.

Lisein, J., Pierrot-Deseilligny, M., Bonnet, S. & Lejeune, P. (2013) A photogrammetric workflow for the creation of a forest canopy height model from small unmanned aerial system imagery. *Forests*, **4** (**4**), 922–944.

Mathews, A.J. & Jensen, J.L. (2012) Three-dimensional building modeling using Structure from Motion: improving model results with telescopic pole aerial photography. In: *Proceedings of 35th Applied Geography Conference*, pp. 10–12. Minneapolis, MN.

Micheletti, N., Chandler, J.H. & Lane, S.N. (2014) Investigating the geomorphological potential of freely available and accessible Structure-from-Motion photogrammetry using a smartphone. *Earth Surface Processes and Landforms*, **40** (**4**), 473–486.

Micheletti, N., Chandler, J.H. & Lane, S.N. (2015) Section 2.2.2: Structure from Motion (SfM-MVS) Photogrammetry. In: L.E. Clarke & J.M. Nield (eds), *Geomorphological Techniques (Online Edition)*. British Society for Geomorphology, London.

Micusik, B. & Pajdla, T. (2006) Structure from Motion with wide circular field of view cameras. *IEEE Transactions on Pattern Analysis and Machine Intelligence*, **28** (7), 1135–1149.

Moreels, P. & Perona, P. (2007) Evaluation of features detectors and descriptors based on 3d objects. *International Journal of Computer Vision*, **73** (**3**), 263–284.

Nebiker, S., Annen, A., Scherrer, M. & Oesch, D. (2008) A light-weight multispectral sensor for micro UAV—Opportunities for very high resolution airborne remote sensing. *The International Archives of the Photogrammetry, Remote Sensing and Spatial Information Sciences*, **37**, 1193–1200.

Neitzel, F. & Klonowski, J. (2011) Mobile 3D mapping with a low-cost UAV system. *The International Archives of the Photogrammetry, Remote Sensing and Spatial Information Sciences*, **38**, 1–6.

Ouédraogo, M.M., Degré, A., Debouche, C. & Lisein, J. (2014) The evaluation of unmanned aerial system-based photogrammetry and terrestrial laser scanning to generate DEMs of agricultural watersheds. *Geomorphology*, **214**, 339–355.

Petrie, G. & Walker, A.S. (2007) Airborne digital imaging technology: a new overview. *The Photogrammetric Record*, **22** (**119**), 203–225.

Plets, G., Gheyle, W., Verhoeven, G. et al. (2012) Three-dimensional recording of archaeological remains in the Altai Mountains. *Antiquity*, **86** (**333**), 1–14.

Quincey, D.J., Bishop, M.P., Kääb, A. et al. (2014) Digital terrain modeling and glacier topographic characterization. In: *Global Land Ice Measurements from Space*, pp. 113–144. Springer, Berlin Heidelberg.

Rasztovits, S. & Dorninger, P. (2013) Comparison of 3D reconstruction services and terrestrial laser scanning for cultural heritage documentation. *ISPRS-International Archives of the Photogrammetry, Remote Sensing and Spatial Information Sciences*, **1** (**2**), 513–518.

Rosnell, T. & Honkavaara, E. (2012) Point cloud generation from aerial image data acquired by a quadrocopter type micro unmanned aerial vehicle and a digital still camera. *Sensors*, **12** (**1**), 453–480.

Rychkov, I., Brasington, J. & Vericat, D. (2013) Computational and methodological aspects of terrestrial surface analysis based on point clouds. *Computers and Geosciences*, **42**, 64–70.

Santagati, C., Inzerillo, L. & Di Paola, F. (2013) Image-based modeling techniques for architectural heritage 3D digitalization: limits and potentialities. *International Archives of the Photogrammetry, Remote Sensing and Spatial Information Sciences*, **5** (**w2**), 555–560.

Shortis, M.R., Bellman, C.J., Robson, S., Johnston, G. J. & Johnson, G.W. (2006) Stability of zoom and fixed lenses used with digital SLR cameras. In: *Proceedings of the ISPRS Commission V Symposium of Image Engineering and Vision Metrology*, September 25–27, 2006, pp. 285–290. Dresden.

Smith, M., Chandler, J. & Rose, J. (2009) High spatial resolution data acquisition for the geosciences: kite aerial photography. *Earth Surface Processes and Landforms*, **34**, 155–161.

Snavely, N., Seitz, S.N. & Szeliski, R. (2006) Photo tourism: exploring photo collections in 3D. *ACM Transactions on Graphics, New York*, **25** (**3**), 835–846.

Stumpf, A., Malet, J.P., Allemand, P., Pierrot-Deseilligny, M. & Skupinski, G. (2015) Ground-based multi-view photogrammetry for the monitoring of landslide deformation and erosion. *Geomorphology*, **231**, 130–145.

Thoeni, K., Giacomini, A., Murtagh, R. & Kniest, E. (2014) A comparison of multi-view 3D reconstruction of a rock wall using several cameras and a laser scanner. In: *Proceedings of ISPRS Technical Commission V Symposium*, pp. 23–25. Riva del Garda, Italy.

Torres, J.C., Arroyo, G., Romo, C. & De Haro, J. (2012) 3D digitisation using Structure from Motion. In: *CEIG - Spanish Computer Graphics Conference*, September 12–14, 2012. Jaén, Spain.

Turner, D., Lucieer, A. & Wallace, L. (2014) Direct georeferencing of ultrahigh-resolution UAV imagery. *IEEE Transactions on Geoscience and Remote Sensing*, **52** (**5**), 2738–2745.

Unger, J., Reich, M. & Heipke, C. (2014) UAV-based photogrammetry: monitoring of a building zone. In: *The International Archives of the Photogrammetry, Remote Sensing and Spatial Information Sciences, Volume XL-5, 2014 ISPRS Technical Commission V Symposium*, June 23–25, 2014. Riva del Garda, Italy.

Verhoeven, G.J., Loenders, J., Vermeulen, F. & Docter, R. (2009) Helikite aerial photography–a versatile means of unmanned, radio controlled, low-altitude aerial archaeology. *Archaeological Prospection*, **16** (**2**), 125–138.

Vericat, D., Brasington, J., Wheaton, J. & Cowie, M. (2009) Accuracy assessment of aerial photographs acquiring using lighter-than-air blimps: low cost tools for mapping river corridors. *River Research and Applications*, **25**, 985–1000.

Vetrella, S., Tripodi, C., Colagiovanni, C. & Alfano, A. (1977) Tethered balloons as geostationary platforms for multispectral radiometry. *Acta Astronautica*, **4** (**5**), 617–624.

Vierling, L.A., Fersdahl, M., Chen, X., Li, Z. & Zimmerman, P. (2006) The Short Wave Aerostat-Mounted Imager (SWAMI): a novel platform for acquiring remotely sensed data from a tethered balloon. *Remote Sensing of Environment*, **103** (**3**), 255–264.

Wang, Q., Wu, L., Chen, S., et al. (2014) Accuracy evaluation of 3D geometry from low-attitude UAV images: a case study at Zijin Mine. In: *ISPRS Technical Commission IV Symposium*, May 14–16, 2014, pp. 297–300. Suzhou, China.

Westoby, M.J., Glasser, N.F., Hambrey, M.J., Brasington, J., Reynolds, J.M. & Hassan, M.A. (2014) Reconstructing historic Glacial Lake Outburst Floods through numerical modelling and geomorphological assessment: extreme events in the Himalaya. *Earth Surface Processes and Landforms*, **39** (**12**), 1675–1692.

Westoby, M.J., Brasington, J., Glasser, N.F. et al. (2015) Numerical modelling of glacial lake outburst floods using physically based dam-breach models. *Earth Surface Dynamics*, **3** (1), 171–199.

Wu, C. (2013) Towards linear time incremental Structure from Motion. In: *IEEE International Conference on 3D Vision-3DV*, June 29–July 1, 2013, pp. 127–134. Seattle, WA.

## Associated Reference

Muñoz-Narciso, E., Béjar, M., Tena, A. et al. (2014) Generación de modelos topográficos a partir de fotogrametria digital automatizada en un río de gravas altamente dinámico. In: S. Schnabel & A. Gómez-Gutierrez (eds), *Avances de la Geomorfología en España 2012-2014. XIII Reunion Nacional de Geomorfologia*, pp. 335–338. Universidad de Extremadura, Càceres.

## Further Reading/Resources

For a more comprehensive analysis of the interpolation methods and determination of accuracy when generating DEMs from 3D data, the reader is referred to chapter 2.3 of the British Society for Geomorphology's online edition of Geomorphological Techniques (Bell 2012). An excellent (and practically focussed) overview of the SfM-MVS method, including software comparisons and tips for successful image acquisition, is given in Micheletti et al. (2015). Colomina and Molina (2014) provide a comprehensive review of the available unmanned aerial systems, including details of regulations and the most popular sensors. Turner et al. (2014), Stumpf et al. (2015), and Ouédraogo et al. (2014) all make interesting comparisons between specific software types.

# 5 Quality Assessment

## Quantifying Error in Structure from Motion-Derived Topographic Data

**Abstract**

As with any emerging technology, a comparison of the ability of Structure from Motion (SfM) to generate high-resolution topographic data with alternative, more conventional methods is a prerequisite before it can be applied confidently to any real-world problem. There has been a recent proliferation of SfM validation studies in the geosciences literature. However, each validation study is subtly different from another, and only when a larger data set is assembled from these individual studies can we learn the true limits of this technique. Yet to date, no such analysis has been performed. A comparison of papers in the geosciences reporting quantitative validation of SfM reveals large variations in (i) validation method, (ii) reference data, (iii) survey platform, (iv) survey range and scale, (v) error metric used for comparison, (vi) terrain under investigation, (vii) software used, and (viii) camera used to generate the imagery. In this chapter, each of these factors is discussed in turn, and existing reports reporting quantitative validation of SfM against other methods are summarised. In addition, this chapter presents, for the first time, a synthesis of these validation studies and calls for greater standardisation in the specific methods applied.

**Keywords**

validation; error; survey range; scale; root-mean-square error

### 5.1 Introduction

This chapter examines the typical errors arising from the production of topographic data sets using the Structure from Motion–Multi-View stereo (SfM-MVS) approach. Several studies have presented quantitative assessments of SfM-derived data sets, by comparing them to other digital surveying data, such as those from differential Global Positioning Systems (dGPS), airborne laser scanning (ALS), total station (TS) or terrestrial laser

*Structure from Motion in the Geosciences*, First Edition. Jonathan L. Carrivick, Mark W. Smith, and Duncan J. Quincey.

Companion Website: www.wiley.com/go/carrivick/structuremotiongeosciences

scanning (TLS) surveys (Chapter 2). Before discussing the applications of SfM-MVS in the geosciences (Chapter 6), the main questions addressed in this chapter are the following: "How accurate can one expect an SfM-MVS topographic model to be?" "What are the main limitations in the application of SfM-MVS to certain problems?" Here we review existing quantitative assessments and develop some of these in detail via illustrative case studies.

This chapter firstly reviews important yet often overlooked nuances of specific validation methods and discusses them in detail. We do this because a number of SfM-MVS validation studies have emerged in the geosciences literature in recent years. Each study lends a unique perspective on the ability of SfM-MVS to represent topography accurately, in that they tend to be focused on different applications, in different environments, at different scales, using different SfM-MVS software (see Chapter 4). They also evaluate the performance of SfM-MVS using different methods and metrics. This chapter represents a synthesis of the results of these recent studies.

## 5.2 Validation Data Sets

SfM-MVS generates dense point clouds sampling surface topography non-selectively. This non-selective nature of the survey is in contrast to the majority of more established digital survey techniques (Chapter 2) where each point is sampled purposefully such as at a break of slope, for example. As SfM-MVS generates fully three-dimensional (3D) point clouds, the most natural validation data set would be derived from TLS because this method also produces non-selectively sampled data in the form of a point cloud. Yet concurrent TLS data are not always available. Moreover, at long ranges, where point spread functions result in relatively large spot sizes and "mixed pixels" in TLS data, there is no *a priori* reason to prefer the TLS data above the SfM-MVS data. An alternative is a comparison of SfM-MVS data with the data from more conventional photogrammetry, though the derived data products are not always the same.

This chapter compiles a database of SfM-MVS validation studies and, as such, requires inclusion criteria to apply to any individual validation study. Given that the primary goal is to evaluate the potential of SfM-MVS as a survey technique in a range of conditions, typically the best optimum survey or surveys are selected from each study. For example, where two surveys using different cameras or different software have been conducted over the same area (e.g. Micheletti et al. 2014; Oúedraogo et al. 2014; Stumpf et al. 2015), the best-performing camera or software is selected. The final database of 50 validation points for quantitative analysis is given in Table 5.1. The effect of these other factors (e.g. cameras and software) is described separately in the text (e.g. Sections 5.9 and 5.10) but is not incorporated into the overall database. Similarly, only sub-aerial (not sub-aqueous) SfM-MVS surveys are included from Woodget et al. (2014). The main focus of the

**Table 5.1** Summary of SfM validation data sets.

| Reference | Validation data | Validation method | Platform | Survey area (m²) | Range (m) | RMSE (m) | MAE (m) | SDE (m) | ME (m) | Terrain |
|---|---|---|---|---|---|---|---|---|---|---|
| James and Robson (2012) | Lab 3D scanner | PP | Ground | 0.01 | 0.7 | 0.0003 | | | | Rock sample |
| Favalli et al. (2012) | TLS | RR* | Ground | 0.31 | 1 | 0.0038 | | | | Rock outcrop |
| Favalli et al. (2012) | TLS | RR* | Ground | 0.17 | 1 | 0.0011 | | | | Rock outcrop |
| Favalli et al. (2012) | TLS | RR* | Ground | 0.15 | 1 | 0.0004 | | | | Rock outcrop |
| Favalli et al. (2012) | TLS | RR* | Ground | 0.083 | 1 | 0.00092 | | | | Stalagmite |
| Favalli et al. (2012) | TLS | RR* | Ground | 0.032 | 1 | 0.00033 | | | | Volcanic bomb |
| Smith et al. (2016) | TLS | PP | Ground | 9 | 2.1 | 0.013 | 0.006 | 0.007 | | Ice surface |
| Smith and Vericat (2015) | TLS | RR | Ground | 8.1 | 5 | 0.009 | 0.007 | 0.007 | 0.006 | Badlands |
| Smith and Vericat (2015) | Total station | PR | Ground | 8.1 | 5 | 0.03 | 0.023 | 0.031 | 0.008 | Badlands |
| Smith and Vericat (2015) | TLS | RR | Ground | 11.5 | 5 | 0.01 | 0.008 | 0.009 | 0.005 | Badlands |
| Smith and Vericat (2015) | TLS | RR | Ground | 20.4 | 5 | 0.019 | 0.014 | 0.019 | 0.000 | Badlands |
| Smith and Vericat (2015) | Total station | PR | Ground | 20.4 | 5 | 0.032 | 0.025 | 0.032 | –0.003 | Badlands |
| Smith and Vericat (2015) | Total station | PR | Ground | 22.4 | 5 | 0.02 | 0.017 | 0.02 | 0.004 | Badlands |
| Smith and Vericat (2015) | TLS | RR | Ground | 22.4 | 5 | 0.01 | 0.007 | 0.01 | 0.003 | Badlands |
| Smith and Vericat (2015) | TLS | RR | Ground | 28.3 | 5 | 0.016 | 0.01 | 0.016 | 0.000 | Badlands |
| Smith and Vericat (2015) | Total station | PR | Ground | 28.3 | 5 | 0.067 | 0.048 | 0.069 | –0.002 | Badlands |
| Thoeni et al. (2014) | TLS | PP | Ground | 105,000 | 7.5 | | 0.006 | 0.008 | | Rock wall |
| Micheletti et al. (2014) | TLS (with ICP) | RR | Ground | 10.2 | 10 | 0.0197 | | | | River bank |
| Ružić et al. (2014) | dGPS | PP | Ground | 3,000 | 10 | 0.07 | 0.06 | | | Rock cliff |
| James and Quinton (2014) | TLS | PP | Ground | 137.5 | 20 | 0.018 | | | | Coastal cliff |
| James and Robson (2012) | TLS | RR | Ground | 150 | 20 | 0.07 | | | | Coastal cliff |
| Fonstad et al. (2013) | dGPS | PP | Helikite | 36,000 | 20 | | | 0.15 | 0.07 | Gravel river |
| Woodget et al. (2014) | Total station | PR | Hexacopter | 2,564 | 25.81 | | | 0.032 | 0.004 | Gravel river |
| Woodget et al. (2014) | Total station | PR | Hexacopter | 2,803 | 26.89 | | | 0.019 | 0.005 | Gravel river |
| Woodget et al. (2014) | dGPS | PR | Hexacopter | 2,084 | 27.53 | | | 0.069 | 0.044 | Gravel river |
| Woodget et al. (2014) | Total station | PR | Hexacopter | 4,382 | 28.39 | | | 0.203 | 0.111 | Gravel river |
| Westoby et al. (2012) | TLS | RR | Ground | 9,000 | 30 | | | | | Hillslope |
| Smith et al. (2014) | dGPS | PR | Ground | 6,500 | 40 | 0.135 | 0.091 | 0.126 | –0.049 | Dry river |

*(Continued)*

**Table 5.1** (*Continued*)

| Reference | Validation data | Validation method | Platform | Survey area ($m^2$) | Range (m) | RMSE (m) | MAE (m) | SDE (m) | ME (m) | Terrain |
|---|---|---|---|---|---|---|---|---|---|---|
| Lucieer et al. (2014) | dGPS | PR | Octocopter | 7,500 | 40 | 0.062 | | 0.062 | | Landslide |
| Harwin and Lucieer (2012) | Total station | PP | Octocopter | 2,000 | 40 | 0.049 | | | | Coast |
| Mancini et al. (2013) | TLS | PP | Hexacopter | 14,400 | 40 | 0.19 | | | 0.05 | Beach dunes |
| Mancini et al. (2013) | dGPS | PR | Hexacopter | 14,400 | 40 | 0.11 | | | −0.01 | Beach dunes |
| Smith and Vericat (2015) | TLS | RR | AutoGyro | 4,710 | 50 | 0.08 | 0.055 | 0.077 | 0.022 | Badlands |
| Smith and Vericat (2015) | Total station | PR | AutoGyro | 4,710 | 50 | 0.0993 | 0.066 | 0.098 | 0.018 | Badlands |
| Oúedraogo et al. (2014) | dGPS | PR | Fixed-wing UAS | 120,000 | 100 | 0.14 | 0.1 | 0.09 | | Agricultural |
| Stumpf et al. (2015) | TLS | PR* | Ground | 40,000 | 110 | 0.056 | 0.031 | | | Landslide |
| Stumpf et al. (2015) | TLS | PR* | Ground | 40,000 | 110 | 0.027 | 0.016 | | | Landslide |
| Stumpf et al. (2015) | TLS | PR* | Ground | 40,000 | 110 | 0.050 | 0.033 | | | Landslide |
| Tonkin et al. (2014) | Total station | PR | Hexacopter | 90,000 | 117 | 0.517 | | | | Moraine |
| Smith and Vericat (2015) | TLS | RR | AutoGyro | 4,710 | 150 | 0.154 | 0.109 | 0.146 | −0.048 | Badlands |
| Smith and Vericat (2015) | Total station | PR | AutoGyro | 4,710 | 150 | 0.182 | 0.121 | 0.181 | −0.020 | Badlands |
| Smith and Vericat (2015) | Total station | PR | AutoGyro | 1,000,000 | 150 | 0.445 | 0.298 | 0.446 | 0.012 | Badlands |
| Smith and Vericat (2015) | TLS | RR | AutoGyro | 4,710 | 250 | 0.374 | 0.208 | 0.349 | −0.133 | Badlands |
| Smith and Vericat (2015) | Total station | PR | AutoGyro | 4,710 | 250 | 0.2793 | 0.181 | 0.269 | −0.076 | Badlands |
| Smith and Vericat (2015) | Total station | PR | AutoGyro | 1,000,000 | 250 | 0.391 | 0.273 | 0.391 | −0.014 | Badlands |
| Vericat (EGU Abstract) | TLS | PR | AutoGyro | | 300 | 0.41 | | | | Gravel river |
| Stumpf et al. (2015) | ALS | PR* | Ground | 105,000 | 500 | 0.204 | 0.143 | | | Landslide |
| Stumpf et al. (2015) | ALS | PR* | Ground | 105,000 | 500 | 0.232 | 0.134 | | | Landslide |
| Javernick et al. (2014) | dGPS | PR | Helicopter | 1,125,000 | 600 | 0.23 | 0.16 | 0.23 | −0.03 | Gravel river |
| James and Robson (2012) | Photogrammetry | RR | AutoGyro | 1,120,000 | 1000 | 1.00 | | | | Volcanic crater |

Points marked * are technically variants of these validation types (see main text).

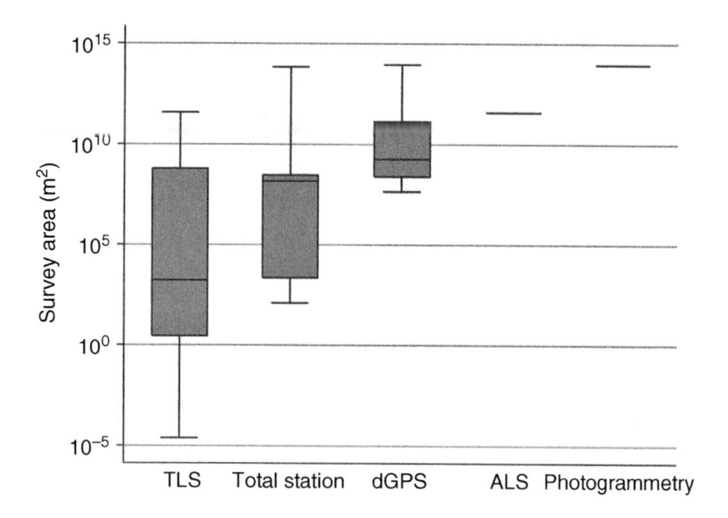


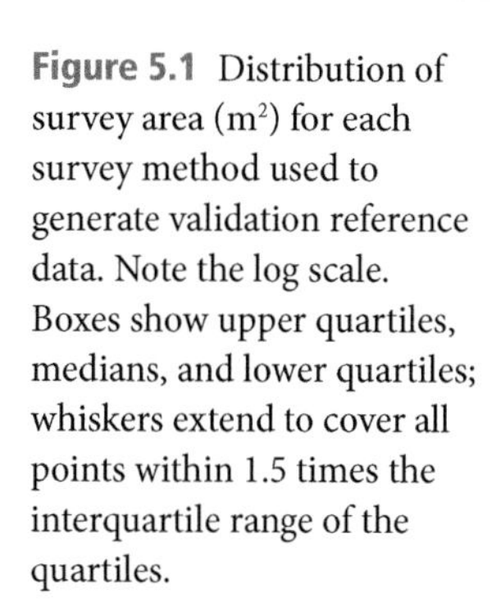
**Figure 5.1** Distribution of survey area ($m^2$) for each survey method used to generate validation reference data. Note the log scale. Boxes show upper quartiles, medians, and lower quartiles; whiskers extend to cover all points within 1.5 times the interquartile range of the quartiles.

discussion in this chapter is on the effect of validation type (Section 5.3), survey platform (Section 5.4), and survey range (i.e. object camera baselines) (Section 5.5) because the greatest variety of data are available to assess these factors quantitatively.

Using the compiled database of 50 SfM-MVS validation data points, Fig. 5.1 examines the distribution of survey area for different reference validation data sets. A few studies have used high-resolution laboratory 3D laser scanners to validate SfM-MVS at very fine scales. James and Robson (2012) and Favalli et al. (2012) examined geological samples which covered only a small area (decimetres). Including the laboratory scanner into a broader TLS classification, TLS data provide the validation data set in 50% of the validation surveys. TLS validation data sets tend to cover a much more extensive area, similar to those for which TS data (28% of the data) are used. dGPS data cover still larger survey areas (8%), while James and Robson (2012) and Stumpf et al. (2015), respectively, used conventional photogrammetry from airborne imagery and ALS to validate a large-scale SfM-MVS-derived topographic model. Figure 5.1 demonstrates clearly that different survey methods are used to validate SfM-MVS data at different scales. In the Section 5.3, the effect of such methodological differences on attempts to draw general conclusions regarding the accuracy and precision of SfM-MVS is considered.

## 5.3 Validation Methods

The critical issue underlying Fig. 5.1 is that different validation studies focus on subtly different aspects of the accuracy and precision of topographic data products. Most fundamentally, the survey technique used to derive the reference data set will determine the specific validation method used and will no doubt influence the results. This effect needs to be analysed in detail before general conclusions can be drawn. A synthesis of extant validation

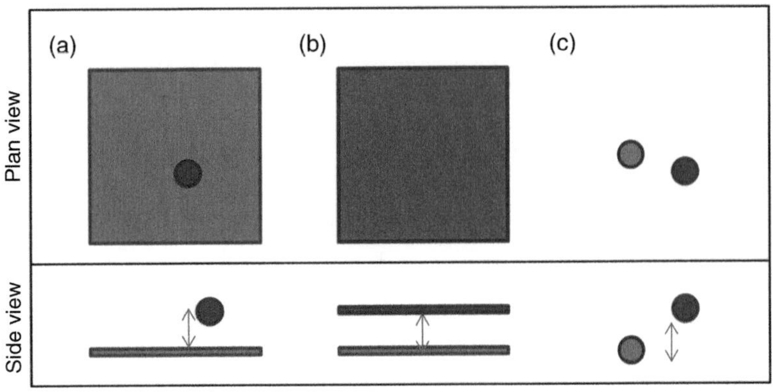

**Figure 5.2** Schematic representation of the three different types of topographic validation methods: (a) point-to-raster, (b) raster-to-raster, and (c) point-to-point.

studies indicates that three common methods are used to compare SfM-MVS with more established surveying methods (Fig. 5.2).

1 **Point-to-raster (PR) comparison** (or more broadly point-to-surface comparison). Since some of the most accurate and precise survey data are provided by TS surveys and dGPS surveys, both of which represent topography as individual points, a large number of studies (e.g. Javernick et al. 2014; Lucieer et al. 2014; Oúedraogo et al. 2014; Smith et al. 2014; Woodget et al. 2014; Smith & Vericat 2015) have used point data from such surveys to validate SfM-MVS-derived data products. However, as SfM-MVS produces point cloud data which are rarely used directly (as 3D points), these validation points are more often validated against rasterised data products derived from SfM-MVS surveys. Typically the mean or minimum of a number of point elevations observed within each grid cell are used as a comparison statistic. The robustness of such validation is determined by the grid size imposed: the smaller the grid size, the smaller the likely error. The range of surface elevation within each grid cell will also influence the result with rougher surfaces exhibiting higher errors than smoother surfaces.

2 **Raster-to-raster (RR) comparison** (or surface-to-surface comparison). Since the vast majority of applications of high-resolution topographic data still require rasterised or 2.5D data products (i.e. digital elevation models or DEMs), several validation studies (e.g. Favalli et al. 2012; James & Robson 2012;Westoby et al. 2012; Micheletti et al. 2014; Smith & Vericat 2015) use such derived data products to validate SfM-MVS. However, this comparison relies on the availability of a survey of a comparable spatial resolution to the SfM-MVS survey, most typically a TLS data set. Where a TLS data set is unavailable, dGPS or TS point data must be interpolated to a uniformly gridded DEM, though this validation method essentially reduces to (1) mentioned earlier. Where TLS validation data are available, a comparison of derived DEMs represents the most appropriate validation of SfM-MVS for

the majority of applications. However, at short ranges, where SfM-MVS point clouds are of a much greater resolution than TLS-derived point clouds, there is no direct justification for automatically preferring TLS data as the "reference data set" over SfM-MVS data, as the errors inherent in the TLS data may be larger than the (admittedly unknown) errors in the SfM-MVS data set.

3 **Point-to-point (PP) comparison.** A small number of validation studies (e.g. James & Robson 2012; James & Quinton 2014) compare point clouds derived from SfM-MVS with points derived directly from TLS. This point cloud approach may naturally appear to be the optimum validation strategy which would yield the lowest error metric. However, it should be acknowledged that (i) except for visualisation purposes point clouds are rarely the final topographic data product, (ii) the specific location of individual points within a point cloud is determined by the pre-specified angular point spacing (for TLS) and the location of keypoints in imagery (for SfM-MVS), and there is no reason to presuppose that these locations will be exactly concordant. Thus, minor offsets in point location will bias such PP comparisons. However, when validation metrics are summarised across entire point clouds, the effects of (ii) are likely to cancel out. Studies which compare dGPS or TS survey points directly with point clouds (e.g. Fonstad et al. 2013) can also be classified in this way, but the limitation outlined in (ii) is even more apparent given the coarseness of the validation data set. CloudCompare (http://www.danielgm.net/cc/) is used regularly to perform such PP comparison; however, for large point clouds this type of comparison becomes computationally demanding.

Variants of the classification given earlier exist in the literature. For example, rather than computing the distance between raster-based DEMs, Favalli et al. (2012) use the mesh comparison tool of Cignoni et al. (1998) to test the distance between two triangular meshes. Stumpf et al. (2015) apply the novel technique of Lague et al. (2013) where surface normals are computed for each point based on all data points within a pre-specified radius. This normal is used to inform the comparison of points within the models. Stumpf et al. (2015) observe that this multi-scale model-to-model cloud comparison (M3C2) technique outperforms similar techniques that compute distances using surface normals (e.g. Girardeau-Montaut et al. 2005).

Aggregating the different survey methods according to the classification of validation type as outlined earlier, Fig. 5.3a examines the usage of each. For ease of analysis, the approach of Favalli et al. (2012) is classified as a variant of RR and that of Stumpf et al. (2015) as a variant of PP, though clearly there are differences. It should be noted that only 26% of studies compare point clouds directly (PP), 34% compare two raster DEMs directly (RR), whilst 40% compare survey points with SfM-MVS-derived rasters (PR). With that in mind, the box plot Fig. 5.3a indicates the typical survey areas to

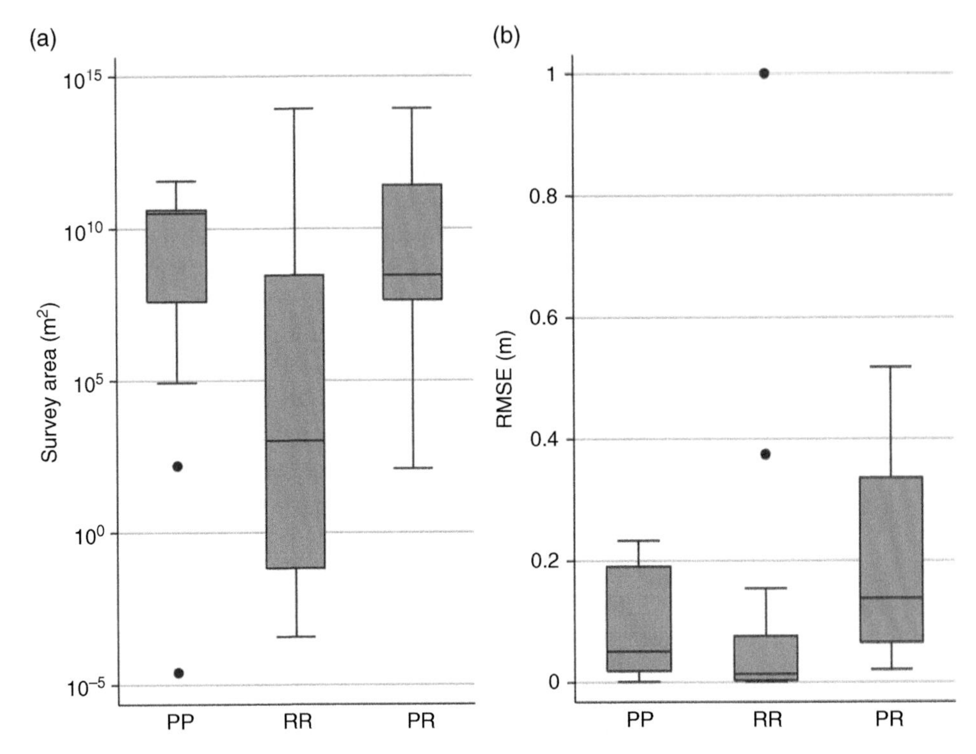


**Figure 5.3** Box plots summarising (a) distribution of survey area (m$^2$) for each validation type and (b) distribution of reported RMSE (m) values for each validation type. PP, point-to-point; RR, raster-to-raster; PR, point-to-raster. See main text for details.

which each method is applied (note the log scale). Figure 5.3b provides an overview of the most commonly reported metric; the root-mean-square error (RMSE) reported in 86% of these validation surveys. There is a clear increase in RMSE for PR validation, with a median RMSE of 50 mm for PP-type validation, 13 mm for RR-type validation, and 137 mm for PR-type validation.

## 5.4 Survey Platform

Imagery for SfM-MVS is collected either from the ground or from the air. Each offers an entirely different perspective of the area under investigation. A little under half (44%) of the validation surveys use aerial imagery (helicopters, multicopters, gyrocopters, fixed-wing unmanned aerial vehicles (UAVs), or helikites) with the remaining 56% utilising ground-based imagery.

Figure 5.4a shows the distributions of reported RMSE error for ground- and aerial-based SfM-MVS surveys, respectively. Mean reported RMSE for aerial surveys (0.277 m) is an order of magnitude greater than that for

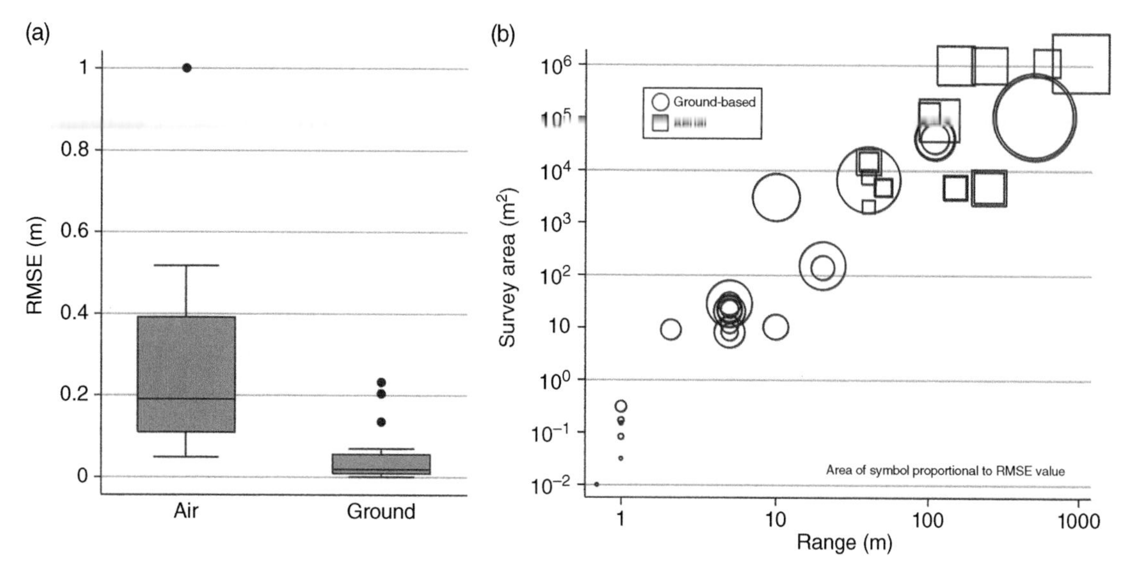


**Figure 5.4** (a) Box plots to show the distribution of RMSE for aerial and ground-based SfM validation surveys. (b) Variability of survey range with survey area for aerial and ground-based SfM studies separately.

ground-based validation surveys (0.043 m). However, this simple comparison masks more fundamental differences in the surveys.

Figure 5.4b plots the survey area against the range at which the survey was conducted separately for aerial and ground-based surveys, where the symbol size is proportional to the RMSE value. Aerial surveys cover a much larger area than ground-based surveys. The largest survey extent for ground-based SfM-MVS validation tests is 9000 $m^2$ (Westoby et al. 2012), which is only slightly higher than the lowest survey area for aerial SfM-MVS validation studies (Fig. 5.4b). The two data sets compiled in Fig. 5.4a do overlap, but there is a noticeable difference in range and area. Moreover, there is a strong correlation between survey area and the range at which the images were taken ($r = 0.70$, $n = 49$). There is a clear increase in RMSE as the survey range and area increases. Therefore, it seems that the observed difference between reported RMSE for ground and aerial SfM-MVS surveys is an artefact of the range at which the survey is conducted.

## 5.5 Survey Range and Scale

Box 5.1 is a case study examining multi-scale validation of SfM-MVS in eroding badlands, which examines the influence of survey range and platform on the quality of the resulting topographic model. Figure 5.5a plots RMSE against survey range for the different validation types described in Section 5.3. From the increase in pixel size with increasing survey range, a linear degradation of precision with survey range might be expected (James &

**Box 5.1** Multi-scale validation of SfM-MVS in eroding badlands

Mark Smith, School of Geography, University of Leeds
Damià Vericat, Department of Environment and Soil Sciences, University of Lleida

### Background and context

One of the main advantages of SfM-MVS is that it can be applied over a variety of scales and from a number of different platforms. How does survey range and platform influence the accuracy and precision of the resulting topographic model? This question is particularly important in the context of monitoring landform change morphometrically, as uncertainties propagate through to the final estimated sediment budget. Here, we present a nested multi-scale validation of SfM-MVS from the plot scale (<30 m²), to the small catchment scale (4710 m²), through to the landscape scale (~1 km²). As reference data sets, we use both point measurements taken using a TS and the most extensive repeat TLS badlands survey to date. This combination enables the assessment of the effect of the specific validation method on the accuracy assessment.

### Method

Here we summarise SfM-MVS validation results from surveys conducted in June 2013 and May 2014. A Leica C10 TLS provided high-resolution reference topographic data for each survey at the small catchment scale. The small catchment was surveyed from 12 stations to minimise and eliminate gaps caused by occlusion. Five plots were also located within this point cloud. Vegetation was removed manually from the point cloud. Point densities were unified using Topographic Point Cloud Analysis Toolkit (ToPCAT; Brasington et al. 2012), providing a 0.1 m resolution DEM. An orthophotograph (Fig. B5.1ia) was used to digitise any areas covered in vegetation, which were eliminated from the comparison. Detrended surface roughness was also computed at this scale. Mean absolute scan registration error was 2 mm. Additional validation points (740) were surveyed at all scales using a TPS1200 TS, averaging 10 consecutive measurements (standard deviation <0.004 mm). Each validation survey was registered to a primary control network of benchmarks in the study area.

SfM-MVS survey details: The five plots were imaged from the ground at 5 m range. At the small catchment scale, four independent sets of images were obtained: (i) an oblique ground survey, (ii) aerial surveys from an AutoGiro gyrocopter at 50 m altitude (AG50 m), (iii) 150 m altitude (AG150 m), and (iv) at 250 m altitude (AG 250 m). The final two surveys were also extended to the landscape scale. Slightly off-vertical imagery was taken from the gyrocopter to minimise doming. A Nikon D310 SLR (focal length 55 mm, 145Mpx) was used. Another small catchment survey was undertaken with a UAV at 50 m altitude in 2013. Photographs were processed in Agisoft Photoscan Professional 1.0.4. Over 100 ground control points (GCPs) were surveyed using a dGPS. These were used to scale and georeference the point cloud and to optimise the

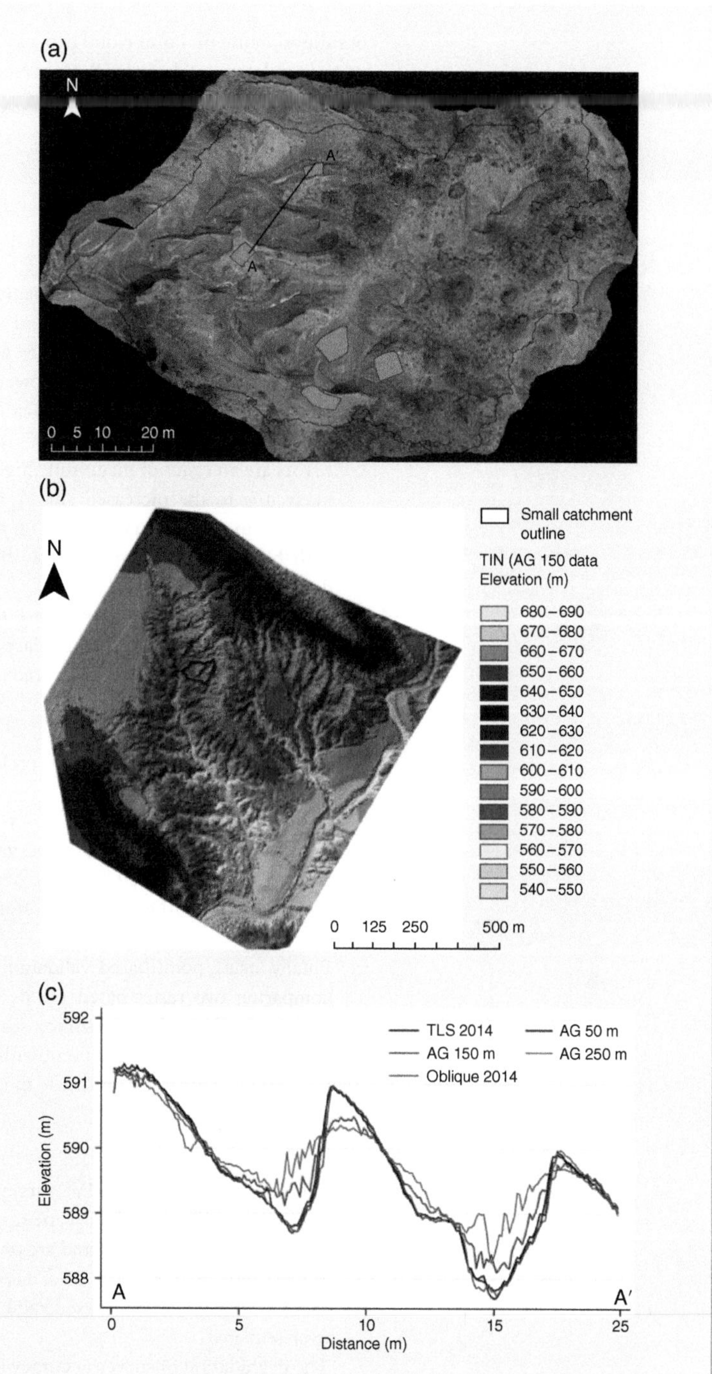


**Figure B5.1i** (a) Orthophotograph of the small catchment (4710 m$^2$) including plot outlines (<30 m$^2$), (b) topographic model of the wider landscape scale (1 km$^2$) study area derived from SfM-MVS, and (c) profiles comparing the TLS DEM with each small catchment scale SfM-MVS DEM. The location of the profile AA' is shown in (a). Source: Smith and Vericat (2015). An interactive example of a point cloud of a badland landscape is available here, courtesy of Mark Smith (see the companion website for the interactive figures).

bundle adjustment. Point cloud processing was undertaken as for the TLS data. At the landscape scale 1 m, DEMs were created owing to the large area under investigation.

### Main findings

Error metrics for both TS and TLS validation are summarised in Fig. B5.1ii. Several points emerge from this:

1. At the plot scale, SfM-MVS is broadly equivalent to TLS and arguably better. RMSE values between the TLS and SfM-MVS on the plots are all less than 0.02 m. Sub-grid roughness values are also comparable (not shown), with SfM-MVS displaying slightly lower roughness, potentially indicating a higher precision (or possibly indicating greater smoothing in the SfM-MVS data).
2. Errors are an order of magnitude higher at the small catchment scale. This is likely due to the increased survey range. Figure B5.1ic clearly shows the effect of increased survey range on the accuracy and precision of the data, with higher-altitude surveys less able to resolve hill crests and valleys in detail.
3. Oblique ground-based surveys are less accurate than aerial imagery from lower altitudes. However, the spatial pattern of errors indicates that over much of the survey, the method is accurate. The higher errors are a result of several slightly misaligned patches (at ~4 m in Fig. B5.1ic), indicating that this is unreliable at large scales.
4. Gyrocopter-based surveys were preferable to UAV (hexacopter) surveys at the same altitude. Images were clearer, and surveys could be controlled better by the user. Clear doming was observed in the UAV SfM-MVS model which was obtained from vertical imagery; such doming was not observed in the SfM-MVS data derived from off-vertical gyrocopter, providing empirical confirmation of the modelling study of James and Robson (2014) (see Box 5.2).
5. Finally, using point-based validation data (Fig. B5.1iia) is less reliable than comparing two raster-based DEMs (Fig. B5.1iib) as the point-based error metrics are dependent on survey scale (see increased error at the landscape scale). Comparing a single point with the mean elevation over a defined grid cell area will tend to overestimate the error in the model.

### Key points for discussion

- Manned gyrocopter SfM-MVS surveys are an inexpensive means of obtaining topographic data at the landscape scale. They can provide data of comparable quality to airborne LiDAR and are preferable to UAV surveys.
- When validating SfM-MVS data, it is important to bear in mind the validation method employed. Point-based validation of DEMs may cause the error to be overestimated.
- The degradation of survey accuracy and precision with survey range is demonstrated clearly.
- SfM-MVS has the potential to produce distributed morphometric sediment budgets of eroding badlands. However, the survey range employed needs to be

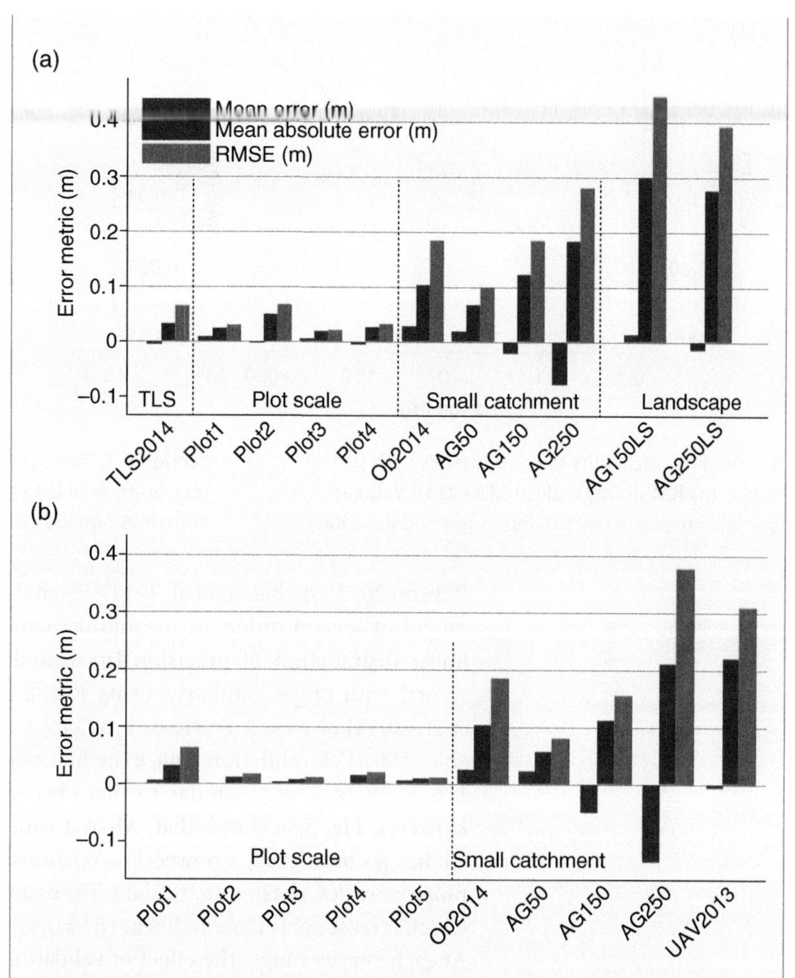


**Figure B5.1ii** Summary of errors in topographic validation at three different scales using (a) total station (TS) data and (b) using terrestrial laser scanning (TLS) data.

given careful consideration. Given the relatively small-magnitude changes expected, survey ranges of greater than 10 m may struggle to resolve real changes reliably. The use of a 10 m inspection pole to capture images may provide the required compromise between survey range and extent.

## Summary

SfM-MVS can produce data comparable to TLS at short ranges; however, a degradation of accuracy with survey range is clearly apparent. This needs to be considered when using SfM-MVS to resolve small changes, in soil erosion studies, for example.

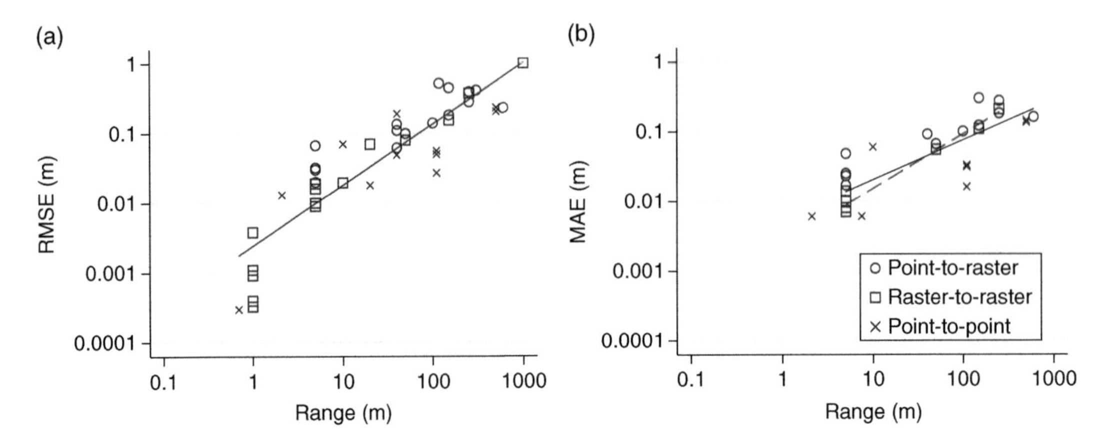


**Figure 5.5** Variability of error metrics with survey range: (a) RMSE (m) and (b) MAE (m). Validation surveys are plotted by validation method described in Section 5.3. Fitted power laws are described in the main text. Sources of data points are described in Table 5.1. Source: Adapted from Smith and Vericat (2015).

Robson 2012; Micheletti et al. 2014). From three data points covering survey ranges of several orders of magnitude, James and Robson (2012) found a linear degradation of precision (measured by the standard deviation of errors) with range. Similarly, using RMSE as the error metric, Micheletti et al. (2014) observed a ratio of 1:625 for two studies. Combining all available SfM-MVS validation data, a median ratio of RMSE and survey range of 1:639 emerges, very similar to that observed by Micheletti et al. (2014). However, Fig. 5.5a shows that, when a wide range of SfM-MVS validation studies is considered, a power–law relationship between RMSE and survey range provides a best fit to the data. The exponent of this relationship is 0.88, which is reasonably close to linear ($R^2 = 0.80$, $n = 43$) (Smith & Vericat 2015). At each survey range, the effect of validation method (PR, RR, and PP) on RMSE can be seen with PR validation methods generally exhibiting the highest errors and PP validation methods exhibiting the lowest errors for a given survey range. No difference is evident between aerial and ground-based SfM-MVS surveys when plotted separately.

The power law fitted in Fig. 5.5a provides a useful summary of errors expected for SfM-MVS surveys at a given range. As seen in Fig. 5.4b, the survey range is somewhat dictated by the study area under investigation, owing to limiting factors associated with survey and post-processing time. At 10 m survey range, approximately 10–15 mm errors are observed. This is more than appropriate for the majority of geoscience applications of SfM-MVS. However, this range falls between that obtainable in ground- and aerial-based SfM-MVS surveys as noted in Fig. 5.4b. The use of an inspection pole to raise a camera up to 10 m above the ground would best achieve this compromise between data extent and survey errors. Remote triggering of the camera facilitates the use of an inspection pole. While the use of an inspection pole has yet to be used in published SfM-MVS validation studies, the analysis in this chapter suggests that it would be an appropriate method of scaling up ground-based SfM-MVS surveys. Indeed, we regard areas as large as 1 ha could feasibly be surveyed in this way.

Considering much larger areas (~1 km$^2$), a minimum survey range of approximately 150 m is required to provide a sufficient coverage without yielding extremely large data sets and increasing processing time (Fig. 5.4b). Errors of approximately 0.1 m should be expected in data sets with large spatial extent, which is similar to errors observed in ALS data sets (see Chapter 2).

## 5.6 Error Metrics

The vast majority of SfM-MVS validation studies (86%) report the RMSE. This is a commonly used error metric in the geoscience literature to compare model predictions with independent observations. Yet, Willmott and Matsuura (2005) note that whilst RMSE represents the error magnitude, it also integrates the distribution of error magnitudes and the square root of the number of errors into its calculation. As a result, Willmott and Matsuura (2005) suggest that the mean absolute error (MAE) represents a more appropriate measure of average error for model performance.

Indeed, MAE is also reported in a large number of SfM-MVS validation studies (56%). Figure 5.5b plots the survey range against MAE for 28 validation studies that reported this metric. A power law best fits this relationship ($R^2 = 0.69$) with a lower exponent of 0.57 (Smith & Vericat 2015). This finding suggests that error as measured by the MAE increases less rapidly, than as measured with RMSE, with increasing survey range. The exponent (0.78) becomes closer to linear, and model fit ($R^2 = 0.97$) improves substantially when just RR validation methods are considered (dashed line in Fig. 5.5b; $n = 8$).

Overall, however, RMSE and MAE are well correlated ($r = 0.993$, $n = 27$) (Fig. 5.6). The standard deviation of error is also reported regularly as a measure of precision. Again, this too correlates well with RMSE ($r = 0.996$, $n = 22$).

Most studies report multiple metrics to give a range of perspectives on model fit (e.g. Javernick et al. 2014; Smith et al. 2014), which also includes mean error. The mean error metric allows positive and negative errors to compensate for each other but provides an insight as to whether or not overprediction or underprediction of surface heights takes place.

Going beyond summary statistics, mapping of the spatial distribution of errors is also common in the literature (e.g. Oúedraogo et al. 2014; Smith & Vericat 2015). This is particularly important as several studies report large differences between different zones of a survey site. They also report clear patterns in errors when mapped spatially (e.g. Favalli et al. 2012; James & Robson 2012; Westoby et al. 2012; Fonstad et al. 2013; Javernick et al. 2014). As detailed in James and Robson (2014), such patterns can provide an insight as to the source of such errors.

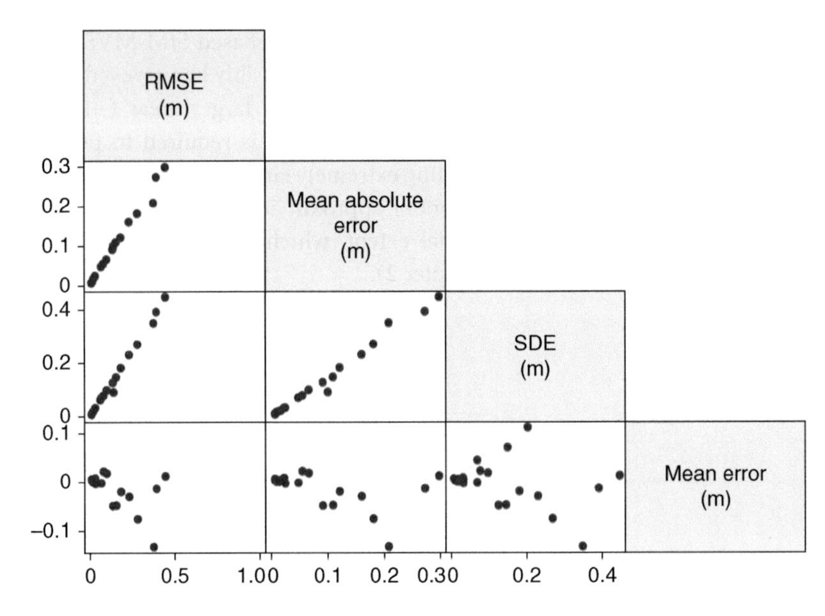


**Figure 5.6** Relationships between error metrics reported in SfM validation studies presented as a scatterplot matrix: root-mean-square error (RMSE), mean absolute error (MAE), standard deviation of error (SDE), and mean error (ME).

Smith and Vericat (2015) also evaluate SfM-MVS performance based on a comparison of sub-grid statistics (i.e. roughness) with coincident TLS data. With multiple applications of surface roughness emerging in the geoscience literature (see Smith 2014 for a review), and with sub-grid roughness being used as an error term when differencing elevation models to determine landform change (e.g. Wheaton et al. 2010; Vericat et al. 2014), such a comparison is particularly relevant. However, any results are specific to the grid size specified.

## 5.7 Distribution of Ground Control Points

As described in Chapter 3, distributing GCPs throughout the survey area is an important part of the SfM-MVS workflow. The distribution of GCPs is discussed in detail in Box 4.1. Two recent studies have examined the effect of clustered GCPs on errors in SfM-MVS-derived topographic models. Extending an aerial SfM-MVS survey reach 1.7 km beyond the extent of ground control on a braided gravel-bed river, Javernick et al. (2014) reported an increase in RMSE from 0.23 to 0.27 m and an increase in MAE from 0.16 to 0.24 m. Similarly, Smith et al. (2014) extended the survey reach upstream 60 m beyond the ground control network which increased the RMSE from 0.14 to 0.47 m and the MAE from 0.09 to 0.25 m because there was a gradual drift in elevation differences with distance from the nearest GCP. James and Robson (2012) also showed that survey errors decreased when GCPs were distributed throughout the survey area. Such investigations highlight the need for robust ground control in SfM-MVS studies; but to date, no published studies have quantified this effect clearly.

GCPs are an integral part of the georeferencing process. Naturally, measurement error in the 3D position of each GCP introduces an additional source of error into the SfM-MVS process as the point cloud is scaled, translated, and rotated into a geographical coordinate system. In some ways, acknowledging this error is important when evaluating the overall accuracy of the SfM-MVS workflow. However, it is often necessary to distinguish between sources of error. This has led some validation studies (e.g. James & Robson 2012; Micheletti et al. 2014) to minimise the separation distance between SfM-MVS-derived point clouds and reference point clouds (typically from TLS) using an Iterative Closest Point algorithm. The Iterative Closest Point algorithm minimises coordinate system alignment errors. While the scale of the point cloud is fixed, the rotation matrix and translation vector are optimised to result in the minimal separation distance between point clouds. Application of Iterative Closest Point algorithms reduces overall reported SfM-MVS errors (e.g. Micheletti et al. 2014) so care should be taken when comparing these "aligned" results with others.

## 5.8 Terrain

The compiled SfM-MVS validation data set covers a wide range of terrain types, including gravel bed rivers, formerly glaciated corries, coastal hillsides, dry ephemeral rivers, recent landslides, eroding badlands, ploughed fields, and volcanic craters (Table 5.1). It is to be anticipated that differences in contrast and landform texture might influence the accuracy of SfM-MVS-derived models, but there is insufficient information available to date to quantify this.

Vegetation provides a particular challenge for SfM-MVS. This challenge arises from the mobility of vegetation in the wind (potentially causing image mismatching), the complexity of vegetation structure, and the requirement for many topographic models to return bare earth points with vegetation filtered out. Most validation data sets are of bare earth elevation (e.g. from TS, dGPS, etc.) though TLS data also provide vegetation surface returns. A number of vegetation filtering algorithms can be applied, from classification of pixels by red–green–blue (RGB) values, using multi-scale dimensionality criteria (Brodu & Lague 2012) to resampling the point cloud at a coarser resolution and extracting the minimum observed elevation over the wider area of dense vegetation (e.g. Javernick et al. 2014).

Vegetation filtering can cause issues with SfM-MVS topographic validation as interpolated surfaces match reference data poorly owing to artefacts arising from the interpolation process. As a result, the majority of validation studies focus on bare, unvegetated surfaces. However, Javernick et al. (2014) compare the accuracy of an SfM-MVS-derived DEM with dGPS points for bare ($n$ = 1985) and vegetated ($n$ = 134) surfaces of a braided river separately. As expected, RMSE is much higher for vegetated surfaces (from 0.17 m over bare areas to 0.78 m in vegetated areas) with similar increases in MAE (from 0.14 to 0.50 m) and standard deviation of error (0.16 to 0.67 m). Tonkin et al. (2014) and Oúedraogo et al. (2014) also report an increase in error with increasing vegetation density.

From aerial data sets and applying basic refraction correction, Woodget et al. (2014) assess the ability of SfM-MVS to obtain topography through water. Validation metrics of exposed surfaces were compared with those of submerged areas for four separate surveys. Average water depths were approximately 0.15 m, though maximum depths of greater than 0.5 m were reported. Once refraction correction had been applied, mean errors increased in only two of the four surveys (0.005 to 0.053 m and 0.004 to −0.008 m) and were lower than exposed areas in the remaining two surveys (from 0.044 to 0.023 m and from 0.111 to −0.029 m).

## 5.9 Software

There have been very few systematic studies to test the performance of different SfM-MVS software in representing topographic data. Chapter 4 details the different software available for SfM-MVS workflows. Whilst some validation studies present only coarse SfM point clouds (e.g. Fonstad et al. 2013), the majority evaluate dense SfM-MVS reconstructions of terrain.

Surveying a ploughed agricultural catchment, Oúedraogo et al. (2014) compared errors from two software packages: Agisoft Photoscan (RMSE = 0.139 m, MAE = 0.1 m) and Mic Mac (RMSE = 0.09 m, MAE = 0.074 m) and suggested that the favourable performance of Mic Mac was due to the camera calibration model as implemented in Agisoft Photoscan. Stumpf et al. (2015) reported similar findings in comparison with VisualSfM. Systematic error arising from the so-called bowl effect was largely removed in Mic Mac as this software implemented Brown's distortion model with five coefficients of radial distortion. James and Robson (2014) suggest that this effect can also be minimised by taking aerial images slightly off-vertical for each flight strip, resulting in convergent imagery (see Box 5.2).

As detailed in Chapter 4, free software is also available for SfM, including Microsoft Photosynth and Autodesk 123D Catch. Micheletti et al. (2014) compared SfM accuracy using 123D Catch (processed online) with that obtainable using Eos Systems Photomodeller. RMSE nearly doubled (from 0.038 to 0.065 m) using 123D Catch, through when an Iterative Closest Point algorithm was applied to align the point clouds, the 123D Catch point cloud produced lower errors than with Eos Systems Photomodeller (from 0.020 to 0.017 m).

Box 5.2 Case study: Minimising systematic error in digital elevation models derived from unmanned aerial vehicle and ground-based imagery

Mike R. James, Lancaster Environment Centre, Lancaster University
Stuart Robson, Department of Civil, Environmental and Geomatic Engineering, University College London

### Background and context

Structure-from-Motion (SfM) algorithms facilitate the use of photogrammetric techniques to derive digital elevation models (DEMs) from imagery acquired with airborne and ground-based consumer cameras. However, they cannot solve issues related to fundamental aspects of the photogrammetric bundle adjustment approach. For UAV surveys, where images are usually acquired with the camera optic axis in near-parallel, vertical orientations, one characteristic error is manifested as a systematic vertical doming deformation of the DEM. Previous work carried out on individual stereo image pairs (Fryer & Mitchell 1987; Wackrow & Chandler 2008, 2011) has demonstrated how self-calibrating bundle adjustment of such parallel image pairs can lead to error in the recovered radial lens distortion and resulting doming deformation. The effect can be also demonstrated in more complex ground-based image strips (Fig. B5.2i), where systematic along-strip deformation occurs when all the images are collected in locally near-parallel directions. If inclined images are also used, systematic deformation is reduced to undetectable levels. Here, we explore how this scales to image blocks typical of aerial surveys and illustrate how the effect can be minimised by the additional inclusion of oblique imagery, or by defining the relationship between doming magnitude and the radial distortion parameter.

### Method

To assess the important parameters behind systematic DEM error, a simulation approach was used in which hypothetical image networks were generated by defining the initial position and orientation of cameras, over a topographic surface represented by a mesh of virtual 3D points. The pixel coordinates at which mesh points could be observed in each image were calculated and small pseudo-random offsets (0.5-pixel standard deviation) added to represent measurement noise. Processing the image network by bundle adjustment allows any resulting systematic DEM deformation to be determined by comparing the coordinates of the adjusted mesh points to their pre-bundle equivalents.

A practical fixed-wing UAV survey was simulated by constructing an image block of two overlapping sets of flight lines (Fig. B5.2ii) with 60% along-strip image overlap and 20% overlap between adjacent strips. The sensitivity of this survey style to doming deformation was considered by exploring the effects of variations in camera height, orientation, and ground slope.

Two approaches to mitigate deformation were then assessed. Firstly, augmenting the image block with oblique images, and secondly, carrying out repeated bundle adjustments with different fixed values of the radial parameter, in order to define the relationship between radial distortion and doming magnitude. More details of the simulation approach can be found in James and Robson (2014).

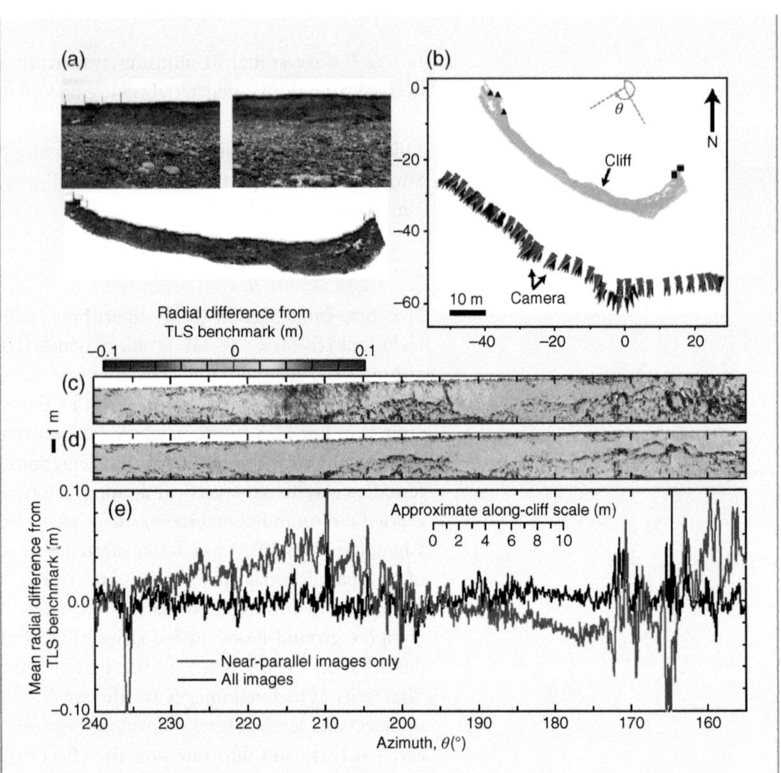


**Figure B5.2i** An elongate cliff model (James & Robson 2012) constructed using photographs taken in a long image strip (a) illustrate how 3D models are prone to systematic error when self-calibration is used and images are collected in locally near-parallel directions (b). Processing only near-parallel cameras, see red in (b), generates a cliff surface that shows systematic deformation (c) when compared with benchmark TLS data (by calculating radial distances from the origin shown in (b)). Negligible systematic error is present when convergent images (shown in black in (b)) are also included in the processing (d). Differences can be highlighted by calculating average errors for approximately 0.1° azimuth segments along the cliff (e). Source: Adapted from James and Robson (2012, 2014).

## Main findings

With parallel viewing directions, self-calibrating bundle adjustment of multi-image networks gives a strongly domed DEM surface (Fig. B5.2iia) in the absence of ground control data. The doming magnitude can be reduced by adding variations to the camera pointing directions or height, with camera pointing direction being significantly more effective than the height. This is in line with previous findings that convergent imagery mitigates systematic doming error in stereo pairs (Wackrow & Chandler 2008, 2011).

The advantages of convergent imagery can be brought to aerial-style image blocks by additionally including a few oblique images (Fig. B5.2iiia,b). However,

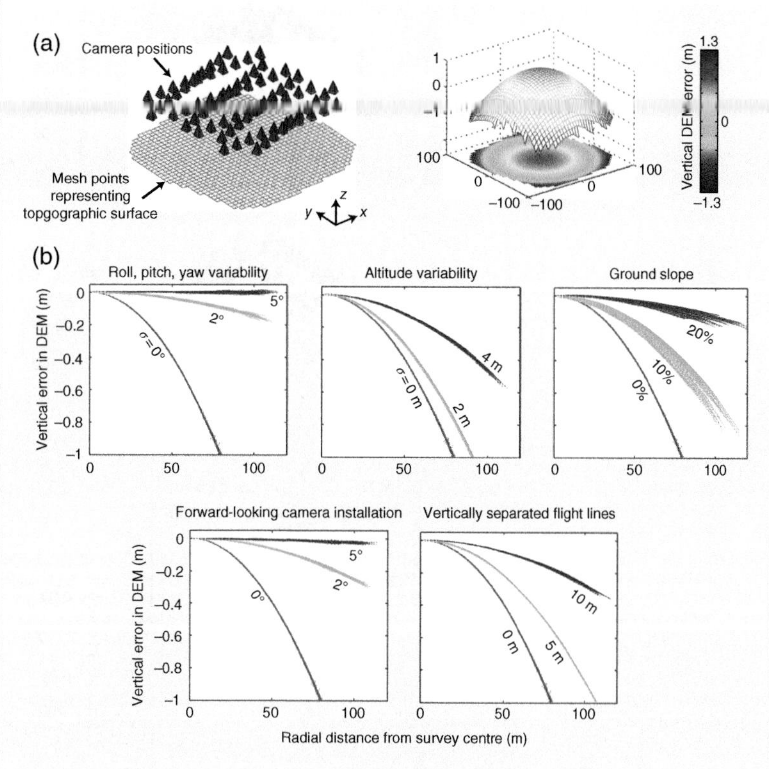


**Figure B5.2ii** Anticipated vertical DEM error in practical image-block scenarios (e.g. for fixed-wing UAVs) processed with self-calibration of radial distortion. (a) With idealised camera positions and orientations over a flat surface, significant systematic DEM doming error results. (b) To facilitate comparisons, systematic vertical DEM error can be represented as error plotted against radial distance from the survey centre, with results translated vertically to give zero at the deformation centre. Upper-row plots illustrate the effect of adding a component of random noise to the camera pointing directions (i.e. variability in UAV roll, pitch, and yaw) or camera altitude, and the effect of surveying over sloping ground. Results are labelled by the standard deviation, σ, of the varied parameter, or the ground slope (in percent). In the lower row, plots demonstrate the effect of non-nadir installation of the camera in the UAV (with the camera forward pointing by the given angle) and by flying the second set of flight lines at an increased altitude (labels give the magnitude of the increases). In all plots, the greatest error is given by the idealised scenario shown in (a) that represents zero noise and flat topography. Source: Adapted from James and Robson (2014).

if this is not possible, a similar level of mitigation can be approached by characterising the relationship between radial distortion parameter and the doming magnitude. With this relationship defined, the radial parameter value associated with minimum DEM distortion can be estimated, and the bundle adjustment is rerun with the parameter fixed at this value (Fig. B5.2iiic).

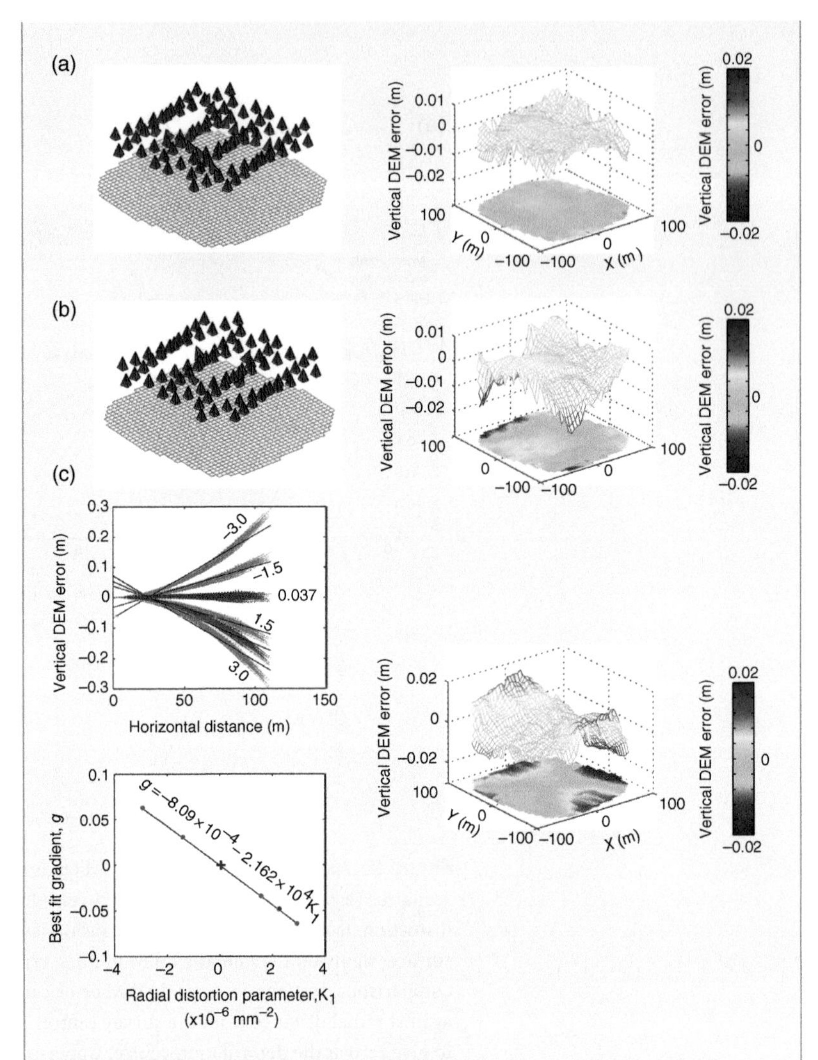


**Figure B5.2iii** Example of practical UAV flight plans to reduce systematic DEM error in self-calibrating image networks through the additional collection of oblique images. (a) Images acquired at 20° inclination to the vertical during two gently banked turns, appropriate for fixed-wing systems. (b) Systems capable of variable camera angles (e.g. some rotor-based UAVs) could minimise additional image capture by capturing fewer, more oblique (30°) overlapping images. (c) If oblique imagery are not available, doming error can be mitigated by deriving a better estimate for the $K_1$ radial distortion parameter. The upper left plot shows the results of using an invariant camera model within the bundle adjustment (grey), with different $K_1$ values (given by the number labels, $\times10^{-6}$ mm$^{-2}$). The red data (with a systematic deformation of up to ~0.2 m in magnitude) result from a self-calibrated image network which recovered a $K_1$ value of $2.2\times10^{-6}$ mm$^{-2}$. The black lines show linear fits to each data set. In the lower left panel, the gradient values for the linear fits demonstrate a linear relationship with $K_1$ (reflecting the correlation between $K_1$ and the surface form) from which the zero-gradient (i.e. minimum doming) $K_1$ value can be estimated (+ symbol, $K_1 = -3.74\times10^{-6}$ mm$^{-2}$). Using this value with an invariant camera model in a bundle adjustment results in a strong mitigation of the doming effect (right panel and blue data in upper left plot). Source: Adapted from James and Robson (2014).

**Key points for discussion**

- We focus here on aerial image networks, but equivalent systematic deformation is likely to occur in ground-based networks if images are taken in near-parallel directions (as shown in Fig. B5.2i). Similar mitigation approaches (convergent imagery) are just as appropriate as for the aerial case.
- The simulations carried out used an "inner constraints" bundle adjustment, that is, without the inclusion of ground control data. The addition of control points into the adjustment will help mitigate the doming effect, but some systematic deformation will remain, at a magnitude accommodated within the uncertainty estimates provided for the control measurements.

**Summary**

If an accurate camera lens distortion model is not available, then image networks with near-parallel viewing directions processed by self-calibrating bundle adjustment will be prone to result in "domed" DEMs, particularly if control is weak or not used. This systematic deformation can be minimised by including oblique images within the self-calibrating bundle adjustment or by better defining radial lens distortion and not using self-calibration within the bundle adjustment.

## 5.10 Camera

A few studies have examined the effect of camera type and model on the quality of SfM-MVS surveys in the geosciences. The effect is complicated by the use of different camera calibration models in different SfM-MVS software. Thoeni et al. (2014) compared the performance of five cameras. Using TLS data for validation, images from a GoPro Hero performed the worst (MAE = 42 mm) followed by those from a camera phone (iPhone 4S; 16 mm). However, consumer grade digital cameras (MAE = 7 mm) performed very similarly to professional grade cameras (MAE = 6 mm). Micheletti et al. (2014) compared the performance of Nikon D700 digital single lens reflex camera (DSLR) (16.2 MP) with that of a camera phone (iPhone 4, 5 MP) and found that they produced similar results at relatively close range. For a close-range survey of an eroding river bank, both DSLR and camera phone models were compared against TLS data by comparing the point clouds directly (PP comparison). Mean error increased only marginally from 6.1 to 8.9 mm though RMSE actually decreased from 38.1 mm with a DSLR to 21.3 mm with a camera phone.

## 5.11 Summary

This chapter has presented an overview of the SfM-MVS validation literature to date. Through assembling and analysing a data set of validation studies, these results have been synthesised to reveal valuable lessons about the ability of SfM-MVS to represent surface topography. In particular, the effect of survey range on precision of the resultant topographic model is clear and appears to be the main limiting factor. However, as summarised in Table 5.1, there are a number of confounding variables that can influence the quality of the topographic model output from SfM-MVS. As noted in Section 5.3, there is no consistent methodology for comparing SfM-MVS-derived topographic data with a reference data set, and the choice of procedure will be determined by the survey method used to derive those data.

More generally, there is an absence of a systematic campaign to validate SfM-MVS robustly on different landforms, using different software, with different cameras, from different ranges and platforms, and with different configurations of GCPs. As a survey method, there is surprisingly little that is standardised in a SfM-MVS workflow. Each operator has a preference as to the following:

- How many pictures are needed
- What resolution they should be at
- The range of angles and perspectives needed
- The degree of overlap in images

These preferences have arisen from personal experience and experimentation. Yet, given the potential impact of SfM-MVS on the scientific community and beyond, and the large number of potential users of the technique, there is a clear need to assemble a validation data set that can offer quantitative insight as to the effect of each of these factors. Thus, the analysis presented here on a relatively small and non-standardised data set is only the first step towards achieving this ambition. With so many interacting factors influencing the quality of the output model, it is difficult to be confident in any attempt to isolate and quantify the effect of any individual factor.

Individual authors (e.g. James & Robson 2012) host images and data online to encourage comparison. Outside of the geoscience literature, the MVS community has assembled a range of high-quality validation data sets that are freely available online at http://vision.middlebury.edu/mview/ (Seitz et al. 2006). The "Middlebury data sets" are used as an objective reference to benchmark different MVS algorithms, the results of which are all available on the website. Anyone is free to submit the results generated from their own code, and standardised protocols are required for result submission. Clearly, implementing such a scheme for the entire SfM-MVS workflow as applied to natural environments is a much larger task. However, hosting image sets online for a range of terrain types for which reference TLS data are available may be a useful starting point in developing a systematic comparison. However, given that many variables arise as part of the image generation process, this provides only a partial solution.

Those issues aside, the synthesis presented herein does provide a meaningful insight into the factors that have the largest effect on the accuracy of

SfM-MVS output, particularly the effect of range on survey precision. Moreover, it is encouraging that at least 35 validation data points have emerged in the geosciences alone within just 3 years (Table 5.1). With improved standardisation of validation protocols, the continued expansion of this database will yield quantitative insights into factors limiting data quality and ultimately provide a clear optimum SfM-MVS workflow for any given survey situation complete with an indication of expected errors.

## References


Brasington, J., Vericat, D. & Rychkov, I. (2012) Modeling river bed morphology, roughness and surface sedimentology using high resolution terrestrial laser scanning. *Water Resources Research*, **48**, W11519. doi: 10.1029/2012WR012223.

Brodu, M. & Lague, D. (2012) 3D t-lidar data classification of complex natural scenes. *ISPRS Journal of Photogrammetric Remote Sensing*, **68**, 121–134.

Cignoni, P., Rocchini, C. & Scopigno, R. (1998) Metro: measuring error on simplified surfaces. *Computer Graphics Forum*, **7** (2), 167–174.

Favalli, M., Fornaciai, A., Isola, I., Tarquini, S. & Nannipieri, L. (2012) Multiview 3D reconstruction in geosciences. *Computers & Geosciences*, **44**, 168–176.

Fonstad, M.A., Dietrich, J.T., Courville, B.C., Jensen, J.L. & Carbonneau, P.E. (2013) Topographic structure from motion: a new development in photogrammetric measurement. *Earth Surface Processes and Landforms*, **38**, 421–430.

Fryer, J.G. & Mitchell, H.L. (1987) Radial distortion and close-range stereophotogrammetry. *Australian Journal of Geodesy, Photogrammetry & Surveying*, **46**, 123–138.

Girardeau-Montaut, D., Roux, M., Marc, R. & Thibault, G. (2005) Change detection on points cloud data acquired with a ground laser scanner. *International Archives of Photogrammetry and Remote Sensing and Spatial Sciences*, **36** (part 3), W19.

Harwin, S. & Lucieer, A. (2012) Assessing the accuracy of georeferenced point clouds produced via multi-view stereopsis from unmanned aerial vehicle (UAV) imagery. *Remote Sensing*, **4** (6), 1573–1599.

James, M.R. & Quinton, J.N. (2014) Ultra-rapid topographic surveying for complex environments: the hand-held mobile laser scanner (HMLS). *Earth Surface Processes and Landforms*, **39**, 138–142.

James, M.R. & Robson, S. (2012) Straightforward reconstruction of 3D surfaces and topography with a camera: accuracy and geoscience application. *Journal of Geophysical Research: Earth Surface*, **117**, F03017. doi: 10.1029/2011JF002289.

James, M.R. & Robson, S. (2014) Mitigating systematic error in topographic models derived from UAV and ground-based image networks. *Earth Surface Processes and Landforms*. doi: 10.1002/esp.3609.

Javernick, L., Brasington, J. & Caruso, B. (2014) Modelling the topography of shallow braided rivers using Structure-from-Motion photogrammetry. *Geomorphology*. doi: 10.1016/j.geomorph.2014.01.006.

Lague, D., Brodu, N. & Leroux, J. (2013) Accurate 3D comparison of complex topography with terrestrial laser scanner: application to the Rangitikei canyon (N-Z). *ISPRS Journal of Photogrammetry and Remote Sensing*, **82**, 10–26.

Lucieer, A., de Jong, S.M. & Turner, D. (2014) Mapping landslide displacements using Structure from Motion (SfM) and image correlation of multi-temporal UAV photography. *Progress in Physical Geography*, **38**, 97–116.

Mancini, F., Dubbini, M., Gattelli, M., Stecchi, F., Fabbri, S. & Gabbianelli, G. (2013) Using unmanned aerial vehicles (UAV) for high-resolution reconstruction of topography: the structure from motion approach on coastal environments. *Remote Sensing*, **5** (12), 6880–6898.

Micheletti, N., Chandler, J.H. & Lane, S.N. (2014) Investigating the geomorphological potential of freely available and accessible Structure-from-Motion photogrammetry using a smartphone. *Earth Surface Processes and Landforms*. doi: 10.1002/esp.3648.

Oúedraogo, M.M., Degré, A., Debouche, C. & Lisein, J. (2014) The evaluation of unmanned aerial systems-based photogrammetry and terrestrial laser scanning to generate DEMs of agricultural watersheds. *Geomorphology*. doi: 10.1016/j.geomorph.2014.02.016.

Ružić, I., Marović, I., Benac, Č. & Ilić, S. (2014) Coastal cliff geometry derived from structure-from-motion photogrammetry at Stara Baška, Krk Island, Croatia. *Geo-Marine Letters*, **34** (6), 555–565.

Seitz, S.M., Curless, B., Diebel, J., Scharstein, D. & Szeliski, R. (2006) A comparison and evaluation of multi-view stereo reconstruction algorithms. In: *Computer Vision and Pattern Recognition, IEEE Computer Society Conference 1*, June 17–22, 2006, pp. 519–528.

Smith, M.W. (2014) Roughness in the Earth Sciences. *Earth-Science Reviews*, **136**, 202–225.

Smith, M.W. & Vericat, D. (2015) From experimental plots to experimental landscapes: topography, erosion and deposition in sub-humid badlands from Structure-from-Motion Photogrammetry. *Earth Surface Processes and Landforms*. doi: 10.1002/esp.3747.

Smith, M.W., Carrivick, J.L., Hooke, J. & Kirkby, M.J. (2014) Reconstructing flash flood magnitudes using "Structure-from-Motion": a rapid assessment tool. *Journal of Hydrology*, **519**, 1914–1927.

Smith, M.W., Quincey, D.J., Dixon, T., Bingham, R.G., Carrivick, J.L., Irvine-Fynn, T.D. and Rippin, D.M., 2016. Aerodynamic roughness of glacial ice surfaces derived from high resolution topographic data. *Journal of Geophysical Research: Earth Surface* doi: 10.1002/2015JF003759.

Stumpf, A., Malet, J.P., Allemand, P., Pierrot-Deseilligny, M. & Skupinski, G. (2015) Ground-based multi-view photogrammetry for the monitoring of landslide deformation and erosion. *Geomorphology*, **231**, 130–145.

Thoeni, K., Giacomini, A., Murtagh, R. & Kniest, E. (2014) A comparison of multi-view 3D reconstruction of a rock wall using several cameras and a laser scanner. In: *Proceedings of ISPRS Technical Commission V Symposium*, June 23–25, 2014, pp. 23–25. Riva del Garda, Italy.

Tonkin, T.N., Midgley, N.G., Graham, D.J. & Labadz, J.C. (2014) The potential of small unmanned aircraft systems and structure-from-motion for topographic surveys: a test of emerging integrated approaches at Cwm Idwal, North Wales. *Geomorphology*. doi: 10.1016/j.geomorph.2014.07.021.

Vericat, D., Smith, M.W. & Brasington, J. (2014) Patterns of topographic change in sub-humid badlands determined by high resolution multi-temporal topographic surveys. *Catena*, **120**, 164–176.

Wackrow, R. & Chandler, J.H. (2008) A convergent image configuration for DEM extraction that minimises the systematic effects caused by an inaccurate lens model. *Photogrammetric Record*, **23**, 6–18.

Wackrow, R. & Chandler, J.H. (2011) Minimising systematic error surfaces in digital elevation models using oblique convergent imagery. *Photogrammetric Record*, **26**, 16–31. doi: 10.1111/j.1477-9730.2011.00623.x.

Westoby, M.J., Brasington, J., Glasser, N.F., Hambrey, M.J. & Reynolds, J.M. (2012) "Structure-from-Motion" photogrammetry: a low-cost, effective tool for geoscience applications. *Geomorphology*, **179**, 300–314.

Wheaton, J.M., Brasington, J., Darby, S.E. & Sear, D.A. (2010) Accounting for uncertainty in DEMs from repeat topographic surveys: improved sediment budgets. *Earth Surface Processes and Landforms*, **35**, 136–156.

Willmott, C.J. & Matsuura, K. (2005) Advantages of the mean absolute error (MAE) over the root mean square error (RMSE) in assessing average model performance. *Climate Research*, **30**, 79–82.

Woodget, A.S., Carbonneau, P.E., Visser, F. & Maddock, I.P. (2014) Quantifying submerged fluvial topography using hyperspatial resolution UAS imagery and structure from motion photogrammetry. *Earth Surface Processes and Landforms*. doi: 10.1002/esp.3613.

## Further Reading/Resources

The "Middlebury data sets" http://vision.middlebury.edu/mview/ (Seitz et al. 2006) provide a useful perspective on SfM validation. Several studies referenced herein focus on the validation of SfM and are recommended for further reading. These include Javernick et al. (2014), Oúedraogo et al. (2014), Micheletti et al. (2014), and Stumpf et al. (2015).

# Current Applications of Structure from Motion in the Geosciences


Abstract

The geosciences are employing Structure from Motion (SfM) to produce orthophotograph mosaics, three-dimensional (3D) point clouds, and digital elevation models (DEMs) of difference. In overview, many geoscience applications of SfM-MVS to date are essentially proof-of-concept studies, but those with robust analyses of error and uncertainty are invaluable to others to ensure scientific rigour and ultimately appreciation of the possibilities offered by an SfM-MVS workflow. SfM-MVS-derived orthophotograph mosaics have been used for mapping and to calculate planform geometry and surface grain-size distribution. SfM-MVS 3D point clouds have (i) aided definition of complex geometry, such as boulders and cliff or gully undercuts; (ii) been used to determine tree biomass, (iii) provided for novel structural analyses of hard rock geology and (iv) enabled automated compositional analyses of soft sediments. DEMs from SfM-MVS processing have been created across a range of spatial scales and at a range of spatial resolutions and compared in 3D quality to other digital survey methods. DEMs have been differenced to detect topographic changes and hence to infer dynamic processes in glacial, fluvial, coastal, hillslope, dryland, volcanic, and shallow underwater environments. Whilst there are now commercial enterprises offering SfM-MVS services, workflows have yet to be fully embraced within industrial applications.



Keywords

Structure from Motion; multi view stereo; orthophotographs; digital elevation model; point cloud; landform analysis


## 6.1 Introduction

Over the past decade, the geosciences have progressively increased its output of publications concerning the application of Structure from Motion–Multi-View Stereo (SfM-MVS) to landform surveys (Chapter 1).


*Structure from Motion in the Geosciences*, First Edition. Jonathan L. Carrivick, Mark W. Smith, and Duncan J. Quincey.



Companion Website: www.wiley.com/go/carrivick/structuremotiongeosciences

This chapter overviews the present use of SfM-MVS in the geosciences. Specifically, it considers how the use of SfM-MVS has expanded in the geosciences from proof-of-concept studies to inclusion of orthophotograph mosaics either to derive red–green–blue (RGB) attributes for three-dimensional (3D) points or to conduct spatial pattern analysis based on colour texture, answering questions about spatial variations in landform properties. It is highlighted that the use of 3D point clouds to date has achieved characterisation of truly 3D landforms such as geological hand specimens, boulders, and undercut cliffs. The most common use of SfM-MVS is as a means to construct **digital elevation models** (DEMs). These are gridded from the 3D point clouds and have been used for mapping, for landform geometry analysis, for surface texture analysis and for input to numerical models. This chapter progresses to show how the geosciences have differenced DEMs to produce a consideration of temporal change, most commonly of landslide surface velocity, glacier surface velocity, and gully erosion rates. Finally, this chapter illustrates the arguably most innovative uses of SfM-MVS in the geosciences where the ability of SfM-MVS to survey at fine resolution *both* spatially and temporally is exploited to address novel research questions. In particular exciting developments are happening where orthophotograph mosaic analysis is fully integrated with 3D point cloud analysis, for example, for geological structure analysis, for automated sediment facies logging and for vegetation canopy biomass estimates.

## 6.2 Use of SfM-MVS-Derived Orthophotograph Mosaics

The SfM-MVS workflow automatically produces georectified and mosaicked images, which may be georeferenced if ground control points (GCPs) are used. However, whilst SfM-MVS-derived orthophotograph mosaics are far quicker to produce than those via conventional photogrammetry, they are presently underused by the geosciences, which have an apparent preference of using SfM-MVS to produce 3D topography in the form of gridded DEMs.

SfM-MVS-derived orthophotograph mosaics are often used for visualisation, as a background image for displaying other data upon. An interactive example of orthophotos visualised as draped over a 3D point cloud mosaic of a reach of the River Cinca can be found on the companion website, courtesy of Damia Vericat. Relatively few studies have included orthophotograph mosaic data and those that have either used them for planimetric mapping or for image analysis to derive landform characteristics. Examples of mapping from SfM-MVS-derived orthophotograph mosaics include Hugenholtz et al. (2013) and Lisein et al. (2013) who identified fine-scale biogeomorphic aeolian landforms and who calculated the surface area of wildlife sampling strips, respectively. The wildlife sampling strips mapping emphasised that for conservation efforts this exercise needed to be done repeatedly and quickly and robustly, so a test between SfM-MVS georeferencing and image footprint projection was made. Lisein et al. (2013) compared both these automatic classifications to field measurements and concluded that the image footprint projection method, in comparison to SfM-MVS, was faster, less demanding in terms of image orientations and overlap, and more accurate in defining wildlife strips. Rippin et al. (2015) used SfM-MVS–derived orthophotograph mosaics to identify both active and relict major meandering supraglacial drainage pathways and a very extensive network of smaller channels in unprecedented spatial detail. They also noted the spatial association of these channels with the structure of the glacier, enabling them to infer processes of supraglacial channel formation and evolution. Casella et al. (2014) mapped maximum wave runup positions from orthophotographs by identifying wet areas and input this information into a numerical model in a fine-resolution analysis of coastal storm activity. Historical aerial and ground-oblique photograph archives have been re-invigorated with the application of SfM-MVS processing to derive orthophotograph mosaics; for example Frankl et al. (2015a) used 9 vertical and 18 low-oblique aerial photographs to map grazing and settlement patterns in the year 1935 over an area of 25 km$^2$ in northern Ethiopia.

**Box 6.1** Case study: Structure from Motion for moraine reconstruction and glacial lake hazard assessment

Matthew Westoby, Geography, Northumbria University

### Background and context

Glacier recession on a global scale has led to an increase in the number and size of proglacial lakes. These lakes are often impounded by sizeable moraine dams, which, if breached, may produce a catastrophic glacial lake outburst flood (GLOF). GLOFs have devastating impacts on downstream communities and are capable of significant geomorphological reworking of the flood path. This case study introduces a selection of SfM-MVS applications for the reconstruction of moraine and floodplain topography to aid the reconstruction of a historic GLOF from Dig Tsho glacial lake (27°52′24.94″N, 86°35′23.60″E) in Sagarmatha (Mt. Everest) National Park, Nepal. The Dig Tsho moraine was breached on August 4, 1985, producing a GLOF that destroyed bridges, property, and valuable agricultural land downstream, as well as causing several fatalities.

### Method

Terrestrial SfM-MVS photogrammetry was used to reconstruct the topography of the breached moraine dam complex (Fig. B6.1i) and an approximately 2 km section of the downstream valley floor. The former permitted calculation of the volume of water drained by the GLOF and the volume of moraine material removed during breach development, as well as the extraction of geometric descriptors of the dam structure for input to a numerical dam breach model (e.g. dam height, crest width, and dam face angles). The latter was used as the topographic domain for two-dimensional (2D) hydrodynamic GLOF modelling.

A total of 2178 photographs of the proglacial area were taken from positions across the terminal moraine and valley flanks using a consumer-grade 12 MP Panasonic DMC-G10 digital camera with automatic focusing and exposure controls enabled. Photographs were divided into several batches to reduce the computational burden of

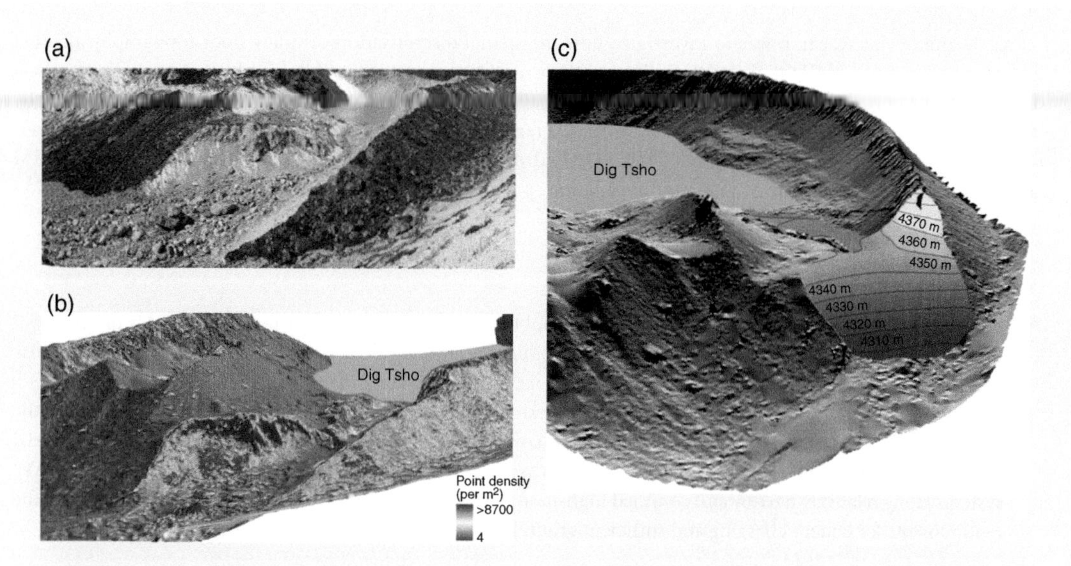


**Figure B6.1i** Panoramic photography of the Dig Tsho terminal moraine breach (a); Point densities (b) across the area visible in (a), and highlighting increased point densities in texturally (and topographically) complex areas within the exposed breach, reconstructed pre-GLOF moraine geometry (c) using manual point cloud editing and surface interpolation techniques (contours for scale).

3D reconstruction, which was performed on the original resolution photographs using SFM-MVS toolkit3 – a freely available open-source application bundle that includes SiftGPU for initial keypoint identification and matching, Bundler for camera pose estimation and sparse point cloud reconstruction, and the MVS algorithms CMVS and PMVS2 for dense point cloud reconstruction.

Dense reconstruction produced $22.6 \times 10^6$ points for the moraine and $6.1 \times 10^6$ points for the floodplain topography. Locally, point densities exceeded $8.7 \times 10^3$ points $m^{-2}$ in texturally complex regions and those photographed at high densities such as exposed moraine material within the dam breach (Fig. B6.1ib). Point cloud sub-sampling was undertaken using the topographic point cloud analysis toolkit (ToPCAT) to reduce the point cloud to a 1 $m^2$ regularised grid, where per-cell elevations represent detrended minimum $z$ values. This additional processing step improved the computational efficiency of subsequent processing steps, including surface reconstruction, and also permitted the extraction of terrain data at resolutions required for 2D GLOF modelling (1–4 $m^2$).

A linear 3D point cloud georegistration was performed in MATLAB and returned residual errors of 1.37, 0.30, and 0.06 m for $xyz$ dimensions for the floodplain DEM, and 0.89, 0.78, and 3.60 m for $xyz$ for the moraine DEM, respectively. High vertical dimension transformation errors for the moraine DEM are attributed to poor target visibility in the dense point cloud.

### Main findings

Using the SfM-MVS-derived DEM the drained volume of the moraine basin was calculated as $5.74 \times 10^6$ $m^3$. The removal of 3D point data describing the internal breach geometry, followed by point interpolation across the breach edges facilitated the reconstruction of an idealised physical representation of the pre-GLOF moraine geometry (Fig. B6.1iii). DEM differencing of "before" and "after" DEMs allowed the calculation of the breach volume, which was returned as $5.81 \times 10^6$ $m^3$ (much lower than a previously published estimate of $9 \times 10^5$ $m^3$). The vast majority of the moraine material removed during breach development was deposited across the first 1–2 km of the floodplain.

A key aim of this research was to investigate how the spatial discretisation of valley floor topography affected patterns and timings of GLOF inundation. Due to their remoteness and inaccessibility, downstream GLOF-prone topography has previously been represented by coarse-resolution terrain models derived from satellite data. The floodplain DEM (Fig. B6.1ii) was decimated to spatial resolutions of 4, 8, 16, and 32 m to mirror a range of levels of topographic detail. Routing of a modelled GLOF hydrograph revealed that progressive coarsening of the DEM altered flow directions and the pattern of inundation, particularly during the early stages of flood propagation (Fig. B6.1iii).

### Key points for discussion

- Terrestrial SfM-MVS photogrammetry is an ideal tool for constructing very high-resolution DEMs of proglacial topography in environments whose remoteness introduces serious logistical impracticalities that preclude the use of cumbersome surveying platforms such as TLS.
- The topography that was reconstructed for this case study only encompassed a 2 km valley reach. To extend the photographic survey range, and therefore the areal coverage of the DEM, the application of low-altitude aerial photography from semi-autonomous platforms such as fixed-wing or multirotor unmanned aerial vehicle (UAV) systems along relatively narrow and confined high-mountain valleys could be explored, but must bear in mind the requirements for battery charging and sufficient ground control.

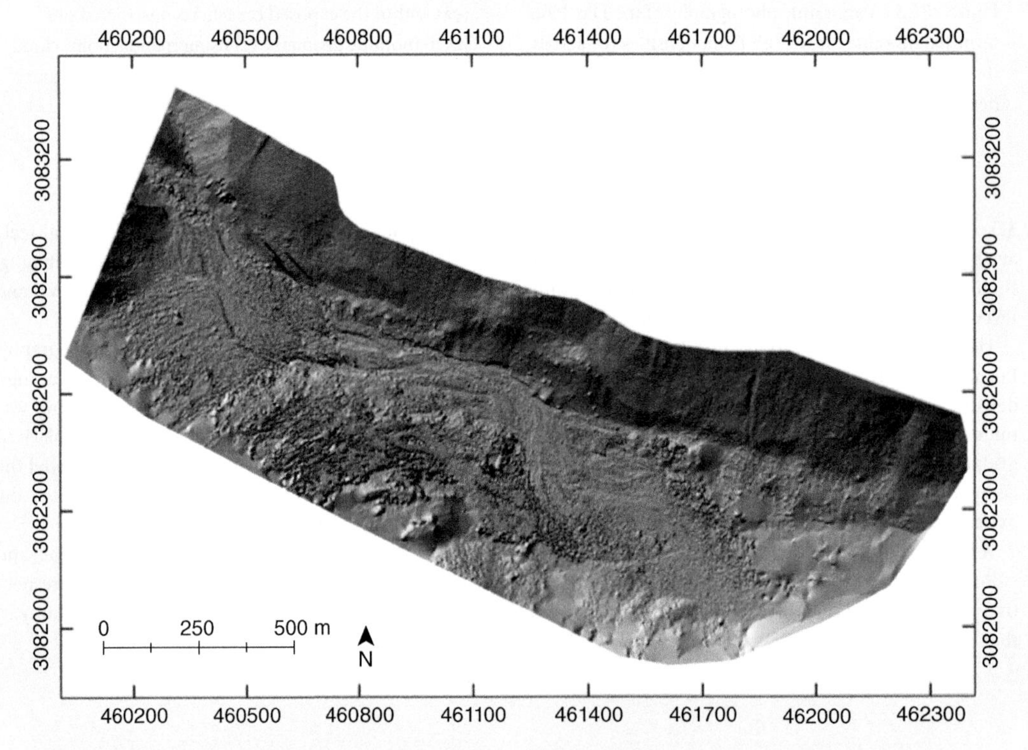


**Figure B6.1ii** Hill-shaded SfM-MVS-derived DEM of the Langmoche Valley (0–2.2 km from breach), produced using terrestrial photography in combination with SfM-MVS toolkit3 for SfM-MVS processing. Source: Data georegistered to UTM zone 45N. Westoby, M. J., Glasser, N. F., Brasington. J., Hambrey, M. J., Quincey, D. J. & Reynolds, J. M. (2014) Modelling outburst floods from moraine-dammed glacial lakes. *Earth-Science Reviews*, **134**, 137–159.

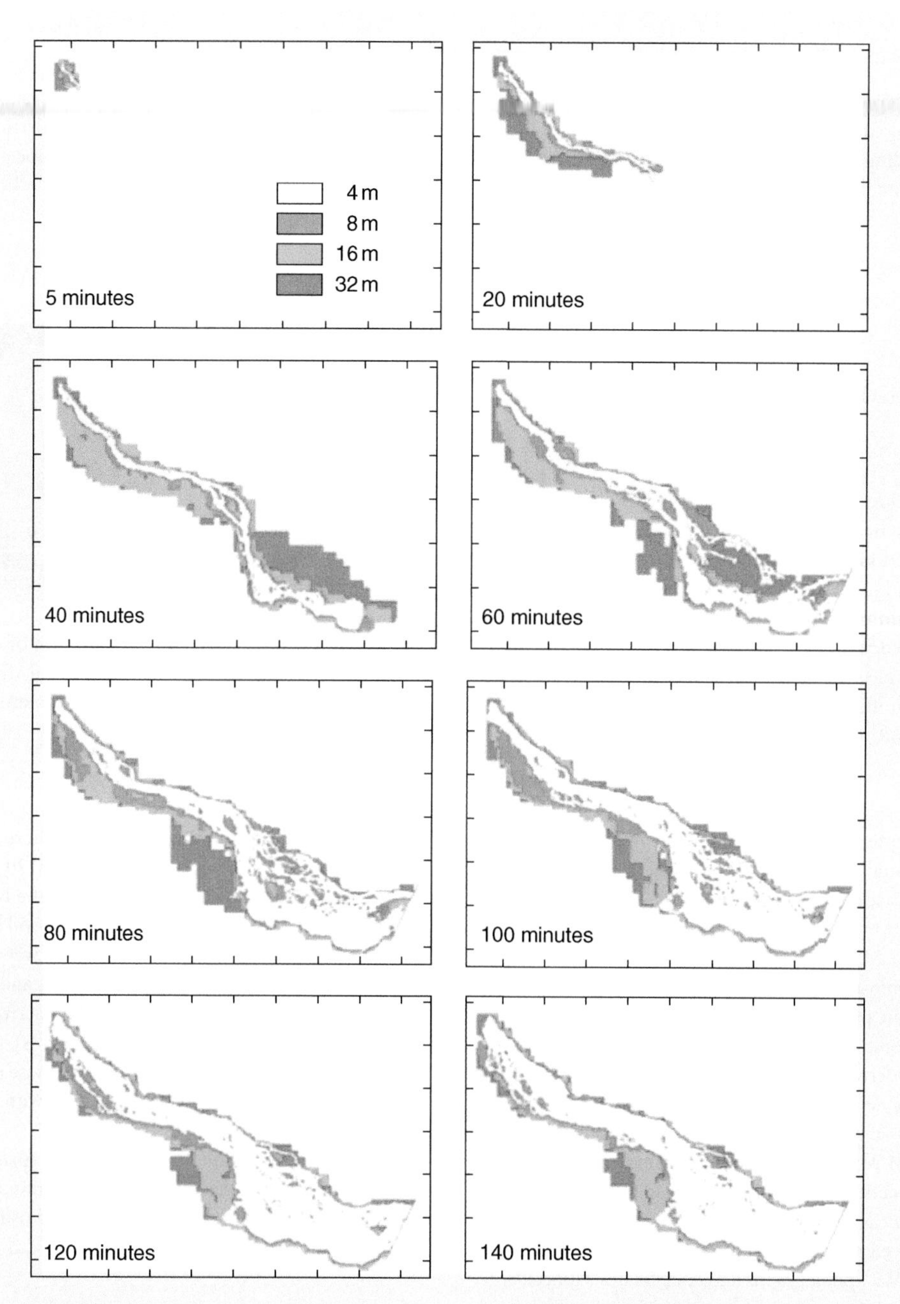


**Figure B6.1iii** Flood inundation at 20-minute intervals for grid resolutions of 4, 8, 16, and 32 m (see "5 minutes" for legend). The 4 and 8 m datasets are broadly similar for all time steps, whereas the use of 16 and 32 m grids results in the inundation of sizeable areas of the valley floor otherwise unaffected by finer grids. Source: Westoby, M. J., Glasser, N. F., Brasington. J., Hambrey, M. J., Quincey, D. J. & Reynolds, J. M. (2014) Modelling outburst floods from moraine-dammed glacial lakes. *Earth-Science Reviews*, **134**, 137–159.

**Summary**

Terrestrial SfM-MVS has demonstrable potential for the reconstruction of moraine and flood path topography which is required as input for the application of advanced numerical and dam breach modelling for GLOF hazard assessment. Extending the downstream coverage, and therefore value, of the SfM-MVS models using semi-autonomous UAVs is an avenue for potential future research.

### Box 6.2 Case study: Reconstruction of flash flood magnitudes

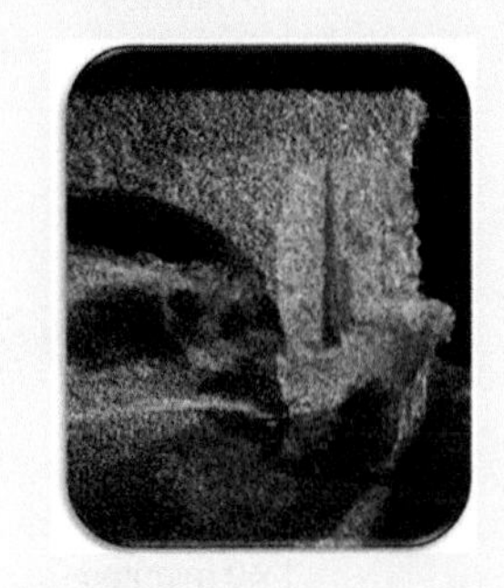

Mark Smith, School of Geography, University of Leeds

#### Background and context

Post-flood analysis is often essential for flash flood magnitude estimation, which in turn is important for flood-frequency analysis and planning. Since flash flood magnitudes are often spatially variable, a distributed survey is needed to characterise an event adequately. Yet, traditional surveys can be time-consuming if high-quality terrain data are to be included. Moreover, a number of uncertainties remain where 1D hydraulic models are applied to cross-sections (e.g. evaluation of channel roughness and superelevation of the water surface around obstacles and bends). The purpose of this study was to demonstrate that SfM-MVS can be used in combination with 2D hydraulic modelling to improve on conventional flash flood reconstructions.

#### Method

On September 28, 2012, widespread flash flooding occurred across Andalucía and Murcia in Spain where 245 mm of rainfall fell in a single morning, causing loss of life and flood damages of around approximately €120 million. This case study details the reconstruction of peak flow magnitude in January 2013 at a single reach: the Barranco del Prado in the Rambla de Torrealvilla. This was part of a broader scale, more conventional post-flood survey.

SfM-MVS was used to provide a detailed DEM of the study reach and to obtain the coordinates of high water marks. This information was incorporated into a hydraulic model (Delft 3D run in depth-averaged 2D mode) and used to estimate peak flow magnitude. The SfM-MVS approach was validated in three stages: (i) the SfM-MVS-derived DEM was compared with differential Global Positioning System (dGPS) topographic validation data (430 points), (ii) SfM-MVS-derived high water marks from desk-based analysis with Agisoft PhotoScan were compared with those obtained from a detailed dGPS field survey, and (iii) discharge estimates for one-dimensional (1D) (cross-sections with uniform flow assumption: slope-conveyance method) and 2D modelling approaches were compared.

**SfM-MVS survey details**: 296 images resampled to 1.2 MP. All images were ground based, taken from surrounding slope crests or channel bed oriented as close to zenithal as possible. Agisoft PhotoScan was used to align images and define camera parameters and produce a dense point cloud (Fig. B6.2i). Georeferencing was performed using GCPs surveyed with dGPS (centimetre-scale residual errors). High water marks were identified manually from multiple SfM-MVS photographs and coordinates exported.

#### Main findings

The multi-step validation of the SfM-MVS approach can be summarised as follows:

1. In comparison with dGPS topographic data, mean absolute errors were less than 0.1 m (when gridded at both 0.1 and 0.5 m resolution). dGPS elevation values are lower than SfM-MVS values. Expanding the comparison upstream (beyond GCPs) reduced the quality of the DEM (errors increased to ~0.25 m). SfM-MVS was better able to represent steep cliffs than manual dGPS surveys.

**Figure B6.2i** Comparison of (a) Structure from Motion generated true colour point cloud and (b) photograph of the study reach looking downstream towards the abandoned aqueduct and the bottom end of the reach. Note: the point cloud in (a) has been cropped to include only the area of interest. Source: Smith et al. (2014).

2 High water marks obtained from Agisoft broadly align with the dGPS high water marks in space and show a similar water surface slope. Mean elevation difference between SfM-MVS-derived high water marks and the nearest dGPS high water mark was 0.016 m, though the mean absolute difference was greater (0.285 m).

3 A 0.5 m SfM-MVS DEM was used in DELFT3D to simulate flow through the reach at different discharges (Fig. B6.2ii). Residuals between high water marks at predicted water surface elevations were minimised to estimate peak discharge. The mean absolute errors for SfM-MVS and dGPS high water marks are minimised at 260 and 280 $m^3 s^{-1}$, respectively. A greater range of values were obtained for the 1D estimates when comparing dGPS and SfM-MVS cross-sections (72–274 $m^3 s^{-1}$).

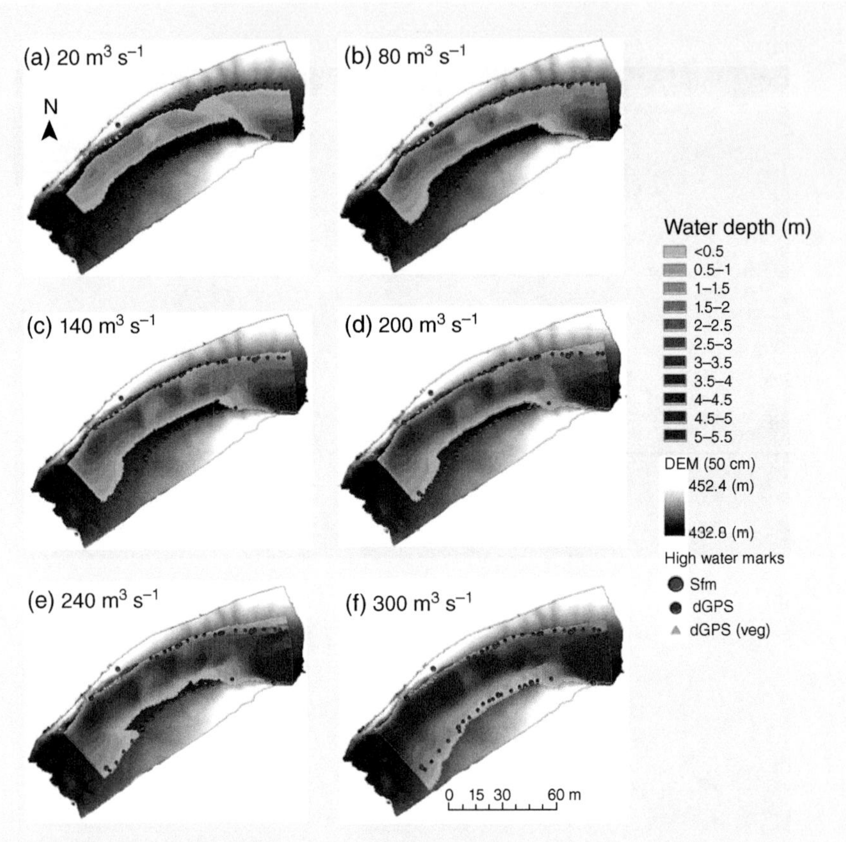


**Figure B6.2ii** Water depth maps for discharges of 20 (a), 80 (b), 140 (c), 200 (d), 240 (e), and 300 (f) $m^3s^{-1}$ as generated with DELFT 3D simulations for different simulated discharges. Surveyed high water marks and the 0.5 m DEM are also displayed. Source: Smith et al. (2014).

### Key points for discussion

- SfM-MVS can be used to speed up post-flood analysis. When coupled with 2D hydraulic modelling, it can reduce uncertainties in peak flood magnitude estimates (by representing form roughness explicitly and simulating cross-channel flow, for example). This represents a robust approach to flash flood discharge estimation.
- Further work could use properties of the DEM to better represent surface conditions in the hydraulic model, by using distributed roughness grids, for example. However, where only the post-flood surface is known, this must be treated with caution.
- High water marks can be extracted from SfM-MVS images directly without the need for detailed field survey. However, a more robust approach would be to highlight these points in the field prior to SfM-MVS survey.
- There is even the potential to crowd source distributed flash-flood surveys in populated areas, though adequate GCPs would be required.
- This method should be used as part of more extensive post-flood analysis methodologies.

### Summary

SfM-MVS offers the potential to both speed up post-flood surveys of flash flood magnitudes and make them more accurate. It is well suited to this application and can increase the reliability of flood-frequency analysis.

Javernick et al. (2014) used image contrast and colour analysis, i.e. standard optical-empirical depth mapping, to estimate water depth from SfM-MVS-derived orthophotographs, thereby creating a seamless data set of above- and below-water topography, for eventual input into a hydrodynamic model. Martín et al. (2013) used orthophotograph mosaics to (manually) map geological structure, and as will be discussed further in Section 6.5 Vasuki et al. (2014) and Gomez (2014) have presented automated mapping based on image analysis for hard rock and for sediments, respectively.

Examples of SfM-MVS-derived orthophotograph mosaics being used for image analysis to derive landform characteristics include analysing variations of brightness values within a local area as an estimate of the surface grain or clast size distribution (de Haas et al. 2014). Feature tracking on pairs of SfM-MVS-derived orthophotograph mosaics has been employed to derive horizontal surface motion, for example, on a glacier (Whitehead & Hugenholtz 2014), on periglacial landforms (Kääb et al. 2013) and on landslides (Niethammer et al. 2012; Lucieer et al. 2014).

## 6.3 Use of SfM-MVS for 3D Point Clouds

The majority of geoscience applications of SfM-MVS are essentially proof-of-concept studies, examining the potential of SfM-MVS as a fast and cheap 3D topographic survey method. These applications span a range of environments and settings and either uses the 3D point cloud or more commonly a gridded derivative in the form of a DEM, the latter of which will be discussed in Section 6.4. Irrespective of whether the final product is the 3D point cloud or a gridded DEM, these studies generally avoid vegetated surfaces. Where vegetation is sparse (enough for some 3D points to intermittently represent bare ground) filtering, usually based on the lowest elevation of a group of points within an area, can remove vegetation automatically (e.g. Javernick et al. 2014).

Studies applying SfM-MVS in the geosciences where the point cloud has been analysed in its own right (i.e. without gridding to a DEM) are usually those concerned with sub-vertical surfaces, that is they exploit the truly 3D properties of the point cloud. Examples include exemplar 3D reconstructions of (i) a decimetre-scale volcanic bomb and a stalagmite by Favalli et al. (2012), (ii) a geological hand sample by James and Robson (2012), and (iii) metre-scale boulders by Gienko and Terry (2014). These studies actually make few measurements from the point clouds, with the exception of volume calculation, perhaps not least because the software for analysing point clouds is very under-developed in comparison to that for grid or raster format data. Additionally, 3D reconstructions of (iv) cliffs by James and Robson (2012), Ružić et al. (2014), Vasuki et al. (2014), and Gomez (2014) have demonstrated the use of SfM-MVS for truly 3D surfaces, including undercuts, over tens of metres, and the same can be said for (v) urban overarching landforms (i.e. bridges) that were reconstructed by Meesuk et al. (2015) for integrating with laser scanner data for improving flood water modelling.

## 6.4 Use of SfM-MVS for Gridded Topography

The majority of geoscience applications of SfM-MVS to date produce 3D topography in the form of a gridded DEM, mainly because this format is at present most easily integrated into a geographical information system (GIS) and lends itself to grid- or raster-based analyses and as input to numerical models. The resultant fine-resolution DEMs can be

- Analysed for landform geometry;
- Inclusive of underwater topography;
- Combined with an orthophotograph mosaic to enable novel analyses.

The first two of these uses are discussed in the subsequent sections and the third is the topic of Section 6.5.

### *6.4.1 Landform Geometry Derived from SfM-MVS*

Use of SfM-MVS-derived DEMs was demonstrated by Westoby et al. (2012; Box 6.1) for determining the dimensions of a breach in a glacial moraine. Other more recent examples are summarised in Table 6.1 but include studies producing planimetric maps and 3D landform geometry measurements including volume of dryland gullies (Castillo et al. 2012; Frankl et al. 2015b). SfM-MVS data has been used for deriving surface microtopography metrics, such as roughness (e.g. de Haas et al. 2014; Leon et al. 2015; Rippin et al. 2015) and reflectance (Rippin et al. 2015). In some cases, the analysis has revealed previously unmapped and unquantified phenomena, for example bioturbation features (Hugenholtz et al. 2013).

### *6.4.2 Bathymetry Derived from SfM-MVS*

Perhaps the first published application of SfM-MVS in the geosciences to sub-aqueous 3D topography was that of a coral reef by Nicosevici and Garcia (2008). Nicosevici and Garcia (2008) were motivated by a need to provide a mapping and positioning underwater tool and their "experiment" or proof of concept used 1100 images obtained on a series of overlapping transects, that when processed through an SfM-MVS workflow produced approximately 160,000 3D vertices and ultimately a dense point cloud across an area approximately $8 \times 8$ m in spatial extent (Fig. 6.1).

Also on a coral reef, Leon et al. (2015) overcame the challenges of shallow (~2 m) water and the associated wave motion, small photographic footprint and thus large ($10^3$) numbers of photos, and variable lighting conditions, to successfully create a DEM at 1 mm resolution along a 250 m transect. They used this fine resolution DEM to extract scale-independent (fractal)

**Table 6.1** Selected applications of SfM-MVS in the geosciences focussed on measurements of 3D surfaces.

| Environment and landform type | Motivation for SfM application | Description of key measurements | Data overview | References |
|---|---|---|---|---|
| Desert alluvial fan | Microtopographic analysis to infer formative processes of terrestrial and Martian fans | Particle size map and surface roughness analysis | ~300 m × 600 m extent of 0.1 m resolution DEM | de Haas et al. (2014) |
| Glacier surface | Very complex microtopography and material properties | Surface roughness and surface reflectance and drainage density | 0.05 m resolution orthophotograph and 0.1 m resolution DEM, both of ~0.25 km$^2$ extent | Rippin et al. (2015) |
| Glacier moraine ridge complex | Difficult-to-access and potentially dangerous terrain | Moraine geometry, especially that of breach | DEM of ~0.1 km$^2$ extent | Westoby et al. (2012) |
| River bed topography and braided river gravel bars, respectively. | Test of accuracy of ground-based SfM-derived 3D data | Differences of DEM grid cell elevation: between SfM and dGPS and ALS surveys and between SfM and dGPS | ~0.04 km$^2$ extent and ~1.2 km$^2$, respectively | Fonstad et al. (2013) and Javernick et al. (2014), respectively |
| Coastal beach, coastal dune system, and coastal cliff, respectively. | Test of accuracy of SfM-derived 3D data | Differences of DEM grid cell elevation: between SfM and TS, between SfM and TLS, and between SfM and dGPS, respectively | ~0.0075 km$^2$, ~0.012 km$^2$, and ~300 m long by ~10 m high cliff, respectively | Harwin and Lucieer (2012), Mancini et al. (2013), and Ružić et al. (2014), respectively |
| Aeolian landforms | Geomorphological mapping at unprecedented resolution | Feature detection and micro-topography analysis to identify vegetation and bioturbation features | 0.01 m resolution orthophotograph mosaic and 0.1 m resolution DEM with ~1 km$^2$ extent | Hugenholtz et al. (2013) |
| Coastal boulders | Volumetric estimates of boulders and thus energy required for movement | Complex clast 3D morphology and irregular 3D land form | 5.9 m$^3$ extent | Gienko and Terry (2014) |
| Dryland channel | Efficient reconstruction of a flash flood | High-water mark 3D positions from SfM-aligned imagery and 3D complex topography | 0.009 km$^2$ extent | Smith et al. (2014) |
| Dryland gully | Test of accuracy of ground-based SfM-derived 3D data | Cross-sectional and volumetric properties of nine gullies | 0.005 km$^2$ each | Castillo et al. (2012) |
| Dryland gully | Understanding of the causes and consequences of soil erosion for sustainable land development | Morphology of four gully heads prone to gully erosion | ~0.01 m resolution DEM over a patch ~5 m × 5 m extent | Frankl et al. (2015b) |
| Coral reef landform | Testing of mapping tool and a positioning system for underwater robot navigation | Different types of typical underwater topologies and textures | 6 m × 12 m patch | Nicosevici and Garcia (2008) |
| Coral reef landform | Benthic and biotope/patch levels (centimetre to metre) roughness | Roughness derived using indices of root-mean-square height, tortuosity, and fractal dimension | 0.001 m resolution orthophotograph and 0.001 m DEM along a 250 m transect | Leon et al. (2015) |

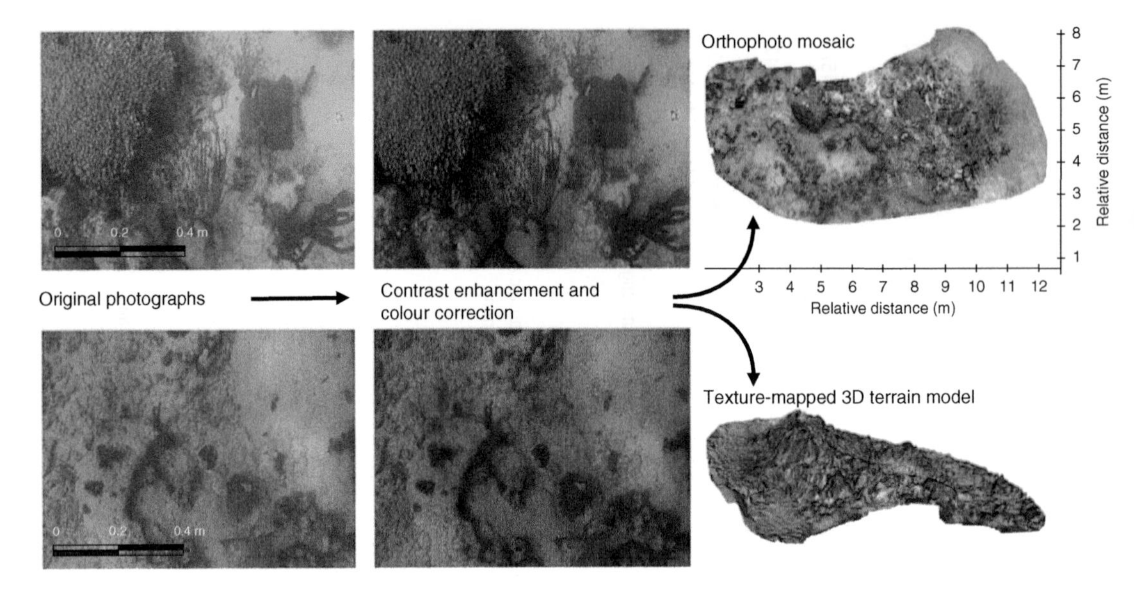


**Figure 6.1** Example of underwater SFM-MVS. Source: Nicosevici, T. & Garcia, R. (2008) Online robust 3D mapping using Structure from Motion cues. In: *OCEANS 2008-MTS/IEEE Kobe Techno-Ocean*, April 2008, pp. 1–7. © IEEE.

roughness, by smoothing (3 × 3 pixel moving window) and detrending (500 × 500 pixel median window), and thereby achieved quantification of scale-independent coral reef roughness (including root mean square height, scale-dependent, and fractal dimension) for the first time at a benthic biotope-relevant scale (centimetre to metre).

Somewhat different in field methods to the underwater surveys of Nicosevici and Garcia (2008) and Leon et al. (2015), and mainly presented as a proof-of-concept workflow, Woodget et al. (2015) employed a UAV to obtain images with through-water visibility. They covered channel lengths of 50–100 m, which they term as mesoscale and thus of an "ecologically meaningful" scale, and generated 3D surfaces at approximately 0.2 m resolution. Notably these 3D surfaces were seamless between terrestrial (above-water) and sub-aqueous settings. One aspect of the novel application of SfM-MVS by Woodget et al. (2015) was that they necessarily corrected the images for refraction, and this processing reduced 3D error to between 0.008 and 0.053 m.

### 6.4.3 Summary of Crossing Spatial Scales

Applications of SfM-MVS in the geosciences have together demonstrated the capability of the SfM-MVS workflow to cross spatial scales. Specifically, SfM-MVS offers seamless surveying across spatial scales, as summarised in Table 6.1 and Table 6.2. There need be no change in the workflow depending on the scale

**Table 6.2** Typical application spatial scales of digital terrain modelling.

| Scale | Point spacing | Typical data source | Example applications |
|---|---|---|---|
| Microscale/surface texture | 0.001–0.1 m | SfM-MVS<br>TLS<br>Lab-based laser scanners | Clast fabric classification<br>Granular movement<br>Surface texture |
| Microscale/surface | 0.1–5.0 m | SfM-MVS<br>dGPS<br>Terrestrial laser scanning | Geotechnical engineering<br>Precision agriculture<br>Mass movement analysis<br>Hydraulic modelling |
| Mesoscale/landform | 5–50 m | SfM-MVS<br>Photogrammetry<br>Airborne laser scanning | Spatial hydrological modelling<br>Spatial soil property analysis<br>Geomorphological mapping |
| Macroscale/landform | 50–200 m | Photogrammetry<br>InSAR<br>Analogue map digitisation | Broad hydrological modelling<br>Biodiversity modelling |
| Macroscale valley | 200 m–5 km | Analogue map digitisation | Environmental lapse rates |
| Macroscale/landscape | 5–500 km | Analogue map digitisation | Global circulation models |

Modified from Hutchinson and Gallant (2000).

of interest, except perhaps the method of camera deployment. The capability of digital survey methods to span several orders of spatial scale has been termed "hyperscale," albeit applied to terrestrial laser scanning (TLS) (Brasington 2010; Brasington et al. 2012; Williams et al. 2014). Arguably SfM-MVS has one of the largest ranges of scale of application of any digital survey method (see Section 2.3.3). Therefore relative to "traditional" classifications of survey spatial scale (e.g. Hutchinson & Gallant 2000), there is now an ability with SfM-MVS to survey (practically) at an unprecedented fine resolution, thus rivalling many laboratory-based devices and very fine resolution outdoor/field scanners.

## 6.5 Combined Orthophotograph and Point Cloud Analysis

Geoscience applications of SfM-MVS that go beyond the derivation of a point cloud, a DEM or a DEM of difference (DoD) to facilitate novel spatial analysis are relatively few but arguably most exciting because these data are enabling novel process understanding. Some of these are via input of

SfM-MVS-derived data to numerical models. For example, Casella et al. (2014) who merged SfM-MVS-derived beach topography with bathymetric datasets and from these extracted topographic transects that were used as an input to a wave runup model. Casella et al. (2014) extracted the position of the maximum wave runup from SfM-MVS-derived orthophotographs and compared this to the numerical model output, using the model to explain *how* the wave runup occurred with respect to spatial variability in landform composition. Similarly, the position of maximum elevation flood marks was identified on SfM-MVS-derived topographic data and used as input to a hydraulic model (Smith et al. 2014; Box 6.2). The ongoing research presented in Box 6.3 is combining image processing and point cloud geometry analysis to determine where (on a reach scale and on a patch scale) bedrock erosion is occurring due to outburst floods (Box 6.3).

The most developed and sophisticated examples of combining orthophotograph analysis and point cloud analysis examples that have been published to date can be categorised as being either concerned with vegetated terrain or with the geological structure and composition of sub-vertical surfaces. Investigation of vegetation spectral dynamics in 3D using SfM-MVS by Dandois and Ellis (2013) is suggested by them to represent a breakthrough in forest ecology. This breakthrough is because despite ecosystem dynamics being analysed with ever more sophisticated remote sensing and field instruments, no single instrument (before SfM-MVS) is technically or logistically capable of combining structural and spectral observations at high temporal and spatial resolutions. Dandois and Ellis (2013) generated 3D point clouds with densities of 30–67 points $m^{-2}$ over a plot 250 m × 250 m from UAV images. They obtained both understory digital terrain models and canopy height models and most novelly mapped RGB values from the orthophotograph mosaics onto the 3D point clouds, thereby enabling the first fine resolution spatiotemporally distributed estimates of above ground biomass and carbon densities.

The use of SfM-MVS-derived data for geological structure and composition of sub-vertical surfaces has been made by Vasuki et al. (2014) for hard rock and by Gomez (2014) for sediments. Both these studies are discussed further in this section. Bemis et al. (2014) have provided a review of the use of SfM-MVS in structural geology, specifically in neotectonics and palaeoseismics. Bemis et al. (2014) reported that neotectonic applications of SfM-MVS are focused on the fundamental measurement of offset features along the length of surface ruptures (e.g. Johnson et al. 2014). Bemis et al. (2014) also demonstrated the utility of orthophotograph mosaics derived from SfM-MVS in palaeoseismics as concerned with increasing automation, and arguably speed and accuracy and objectivity, in representation of exposed stratigraphy and faulting.

Rapid geological mapping, specifically fault and fracture extraction, has been automated using combined image analysis of a georectified orthophotograph mosaic and of a 3D point cloud DEM (e.g. Vasuki et al. 2014; Stumpf et al. 2015). The benefits of this SfM-MVS-based approach

## Box 6.3 Case study: Analysis of bedrock gorge topography and structure

Jonathan Carrivick and Duncan Quincey, School of Geography, University of Leeds
Mingfu Guan, Civil Engineering, University of Leeds

### Background and context

Glacier outburst floods or "jökulhlaups" from either intense ice sheet melt, or from sudden drainage of glacier lakes, affect local populations in West Greenland, the most notable being the town of Kangerlussuaq. In July 2012, floods caused the airport to shut, the bridge to be damaged, the water supply to be interrupted, and loss of mechanical excavator with a financial worth equivalent to a good proportion of the Kangerlussuaq annual budget. The floods route through a series of bedrock gorges, and these are very poorly constrained in existing digital elevation models, which are crucial for understanding flood propagation and hydraulics. This application of SfM-MVS aimed to provide high-resolution DEMs of three gorges for improving flood propagation modelling, but opportunistically has started to examine bedrock erosion.

### Method

At each of three gorges GCPs were set up, comprising natural features visible from as wide a range of positions as possible and with suitable spatial distribution across the area of interest; typically each gorge site was 300 m down valley and 100 m across valley (Fig. B6.3i). Each GCP was marked with biodegradable spray paint and positioned with a dGPS in static mode, average of 60–120 observations depending on radio signal coverage and satellite geometry. Oblique photographs were taken of the entire gorge from both banks and from ground-based and aerial (via quadcopter UAV) positions.

Typically each gorge was imaged with 250 photographs, and these were processed using Agisoft PhotoScan Professional and followed a conventional SfM-MVS workflow. Images were first roughly aligned to establish initial estimations of camera positions and attitude and to generate sparse point clouds on the order of $10^3$–$10^4$ points. We then used a moderate depth filter to derive dense clouds containing $10^6$–$10^7$ points and cleaned the resulting data by manually removing obvious blunders. Each gorge (Fig. B6.3ib) took around 2 days (16 hours) of processing time. The largest amount of computational time was spent extracting the dense point clouds (~8 hours), while overall computational time equated to around 13 of the 16 hours. The point clouds were scaled and georeferenced using ground control data acquired in the field, then decimated and converted to 2D surfaces using the TopCAT (Brasington et al. 2012; Rychkov et al. 2012). Residual error data (i.e. the difference between source values (input data in ground control pane) and estimated values calculated by PhotoScan) were recorded in each case.

Additionally and opportunistically to examine bedrock erosion, 15 patches approximately of 5 m × 5 m surfaces were generated from image sets ranging in size from 44 to 89 photographs and the total number of points in each cleaned dense cloud ranged from 1.16 to $41.26 \times 10^6$. Georeferencing errors ranged from 0.012 to 0.031 m (Fig. B6.3ii). This centimetre accuracy is consistent with reported accuracies for DTMs derived using hand-held consumer-grade sensors and close-range photogrammetry. Each patch took around 1 day (8 hours) to process from the initial camera alignment through to presenting a final, filtered, and geo-referenced point cloud. Most of the computational time was spent in deriving the dense point cloud (~3 hours), and overall computational time was of the order of 5 hours using a 2.8 GHz Intel Core I5 processor and 2 × 2 Gb of RAM.

### Key points for discussion

We anticipate that the exact location of erosion will be largely determined by topography; leading edges are more susceptible to bombardment by stones carried in the flood, and also by geological structure. The automatic detection of geological structure from remotely sensed images has been the focus of a number of previous studies, albeit mostly based on satellite or aerial photograph imagery. Edge detection methods have traditionally focussed on identifying sudden changes

**Figure B6.3i** View upstream from the centre of flood channel in reality (a) and in 3D reconstruction (b). The upper reach of this view is a approximately 10 m high waterfall over which normal flow travels. In times of flood, both banks are inundated, up to the vegetation line at peak flow and with width approximately 100 m.

in image intensity (e.g. Canny, Sobel, Prewitt, Robert, and Laplacian filters), that is, using image orthophotographs rather than digital elevation data. These methods are very effective at demarcating banding and foliation within bedrock surfaces, and also for highlighting fault and fracture surfaces. They are less effective, however, where there is no spectral signature associated with the sudden change in topography. Patch 2 is an excellent example of this, with the perspective view clearly showing a change in topography, but the orthophotograph largely failing to depict this spectrally (Fig. B6.3iii).

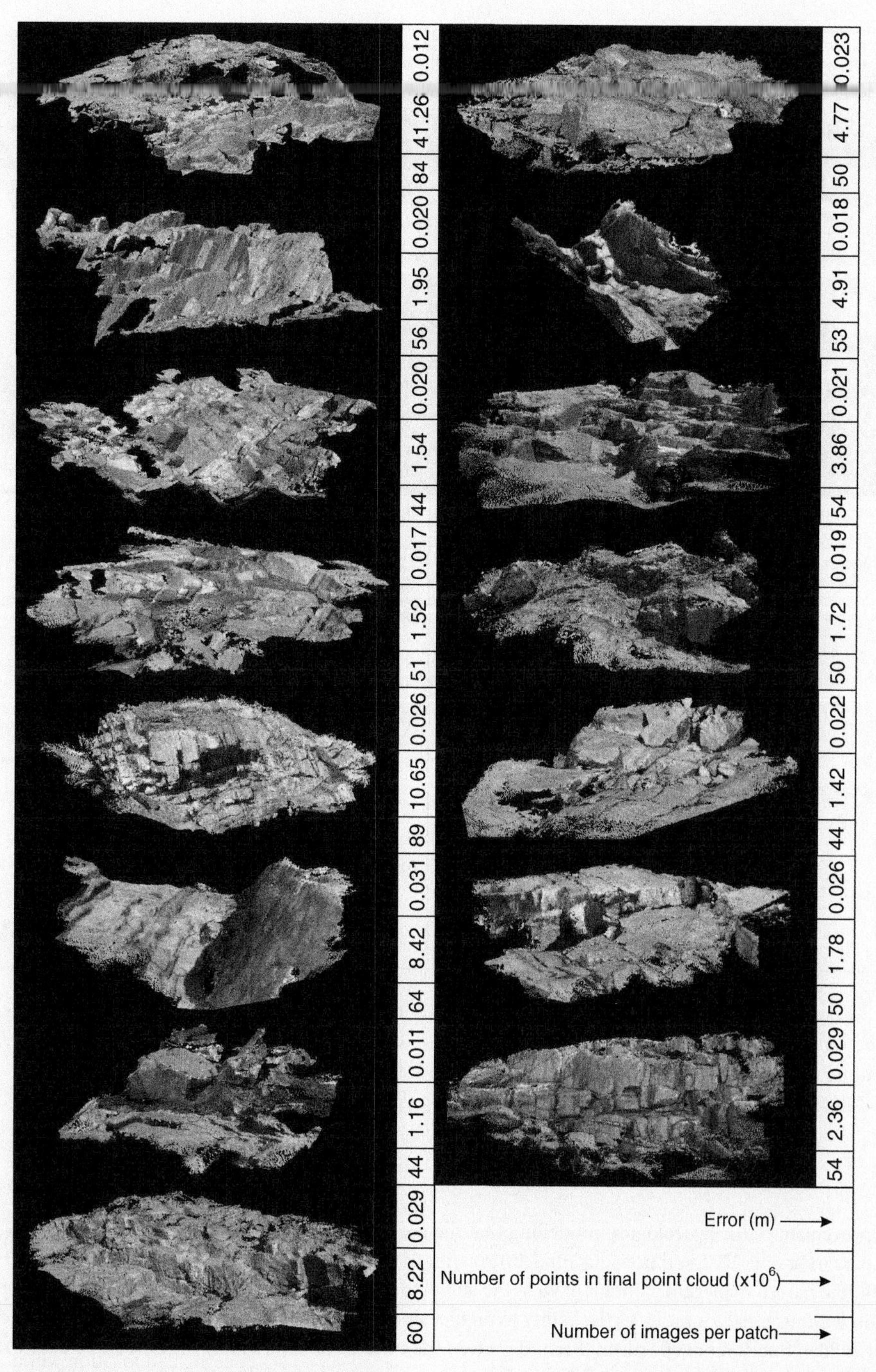


**Figure B6.3ii** Perspective views of the 15 patches, each approximately 5 m × 5 m, surveyed along the flood route. Numerical data to the right of each patch denotes residual error (reported by Agisoft PhotScan), the number of points comprising the final point cloud, and the number of images used in processing the patch. For scale, each patch is approximately 5 m across in its longest dimension. An interactive example of one of these patch point clouds is available on the companion website, courtesy of Duncan Quincey.

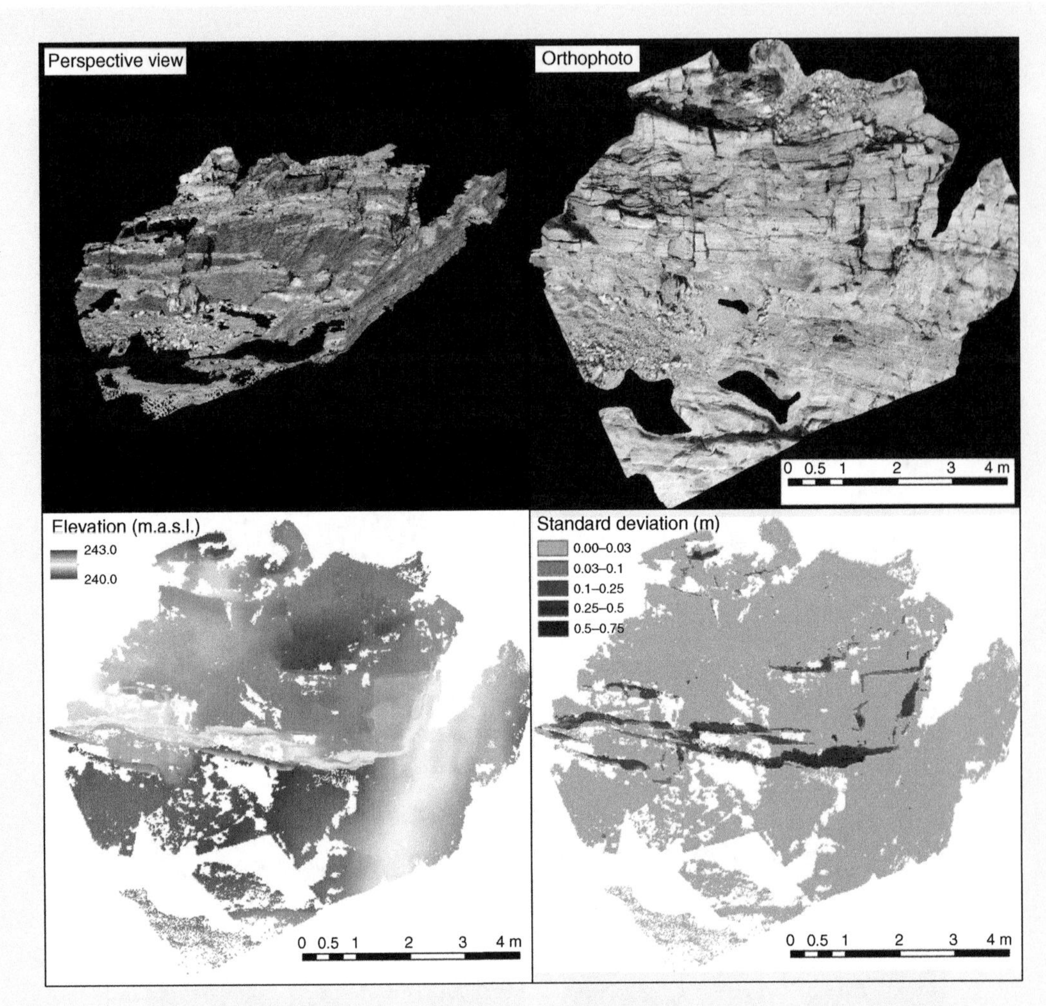


**Figure B6.3iii** Patch 2 in perspective view, orthophotograph, decimated to 2D DEM data, and with leading topographic edges highlighted using the standard deviation of points within each decimated cell as a proxy for relief. Note there is no spectral change in the orthophotograph to depict areas of high and low elevation, as is obvious in the perspective view.

### Main findings

The improvement to the hydrological modelling that the use of a fine-resolution DEM can make becomes apparent when comparing our DEM and previous model runs with those in the current study (Figs B3.6iv and B6.3v). From a scientific and hazard management point of view, the ability to model extreme flows in such fine detail is important for two main reasons. Firstly, it provides the ability to predict areas of inundation with much greater confidence and prepare for sudden floods more effectively. Secondly, given the extra detail afforded by the SfM-MVS data, the ability to include sediment transfer processes into the modelling becomes a real possibility. Previous models have neglected or over-simplified sediment transport and thus failed to provide accurate representations of reality.

Bank and bedrock erosion due to outburst floods is rarely measured, very poorly understood, and largely ignored in outburst flood models to date. Part of the problem is the inaccessibility of the terrain and the necessity for

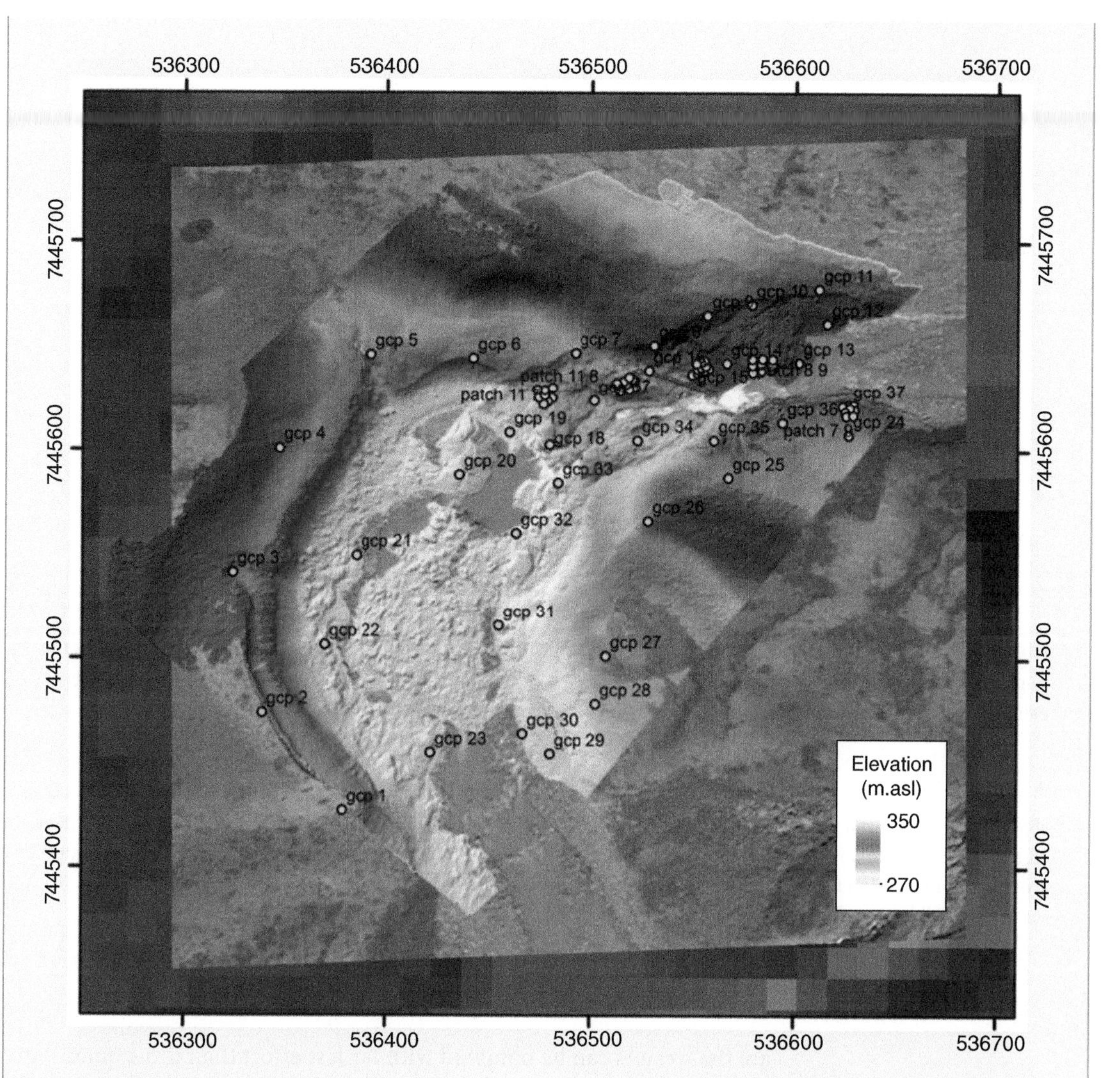


**Figure B6.3iv** Detail of gorge one indicating the distribution of ground control points for the gorge survey and for each of six bedrock patches. The coloured shades denote surface elevation in the final DEM produced by the SfM-MVS workflow, the 0.5 m grid resolution of which contrasts with the 5 m grid cell resolution of the best available DEM from photogrammetry as displayed as a grey shade background. Grid coordinates are projected in UTM22N and are in metres.

repeated surveys to detect change, which often rely on traditional point-based surveying methods and can thus be time consuming and expensive to acquire. Our 15 patch-scale surfaces were derived at the cost of travel, subsistence, and a cheap (< £300) consumer-grade digital camera and can easily be replicated following future floods for comparative analyses.

## Summary

We surveyed three reaches of the flood channel and 15 bedrock patches, and with ground control points acquired using dGPS this took 3 days only. The reach point clouds were used as primary input to a novel computational fluid

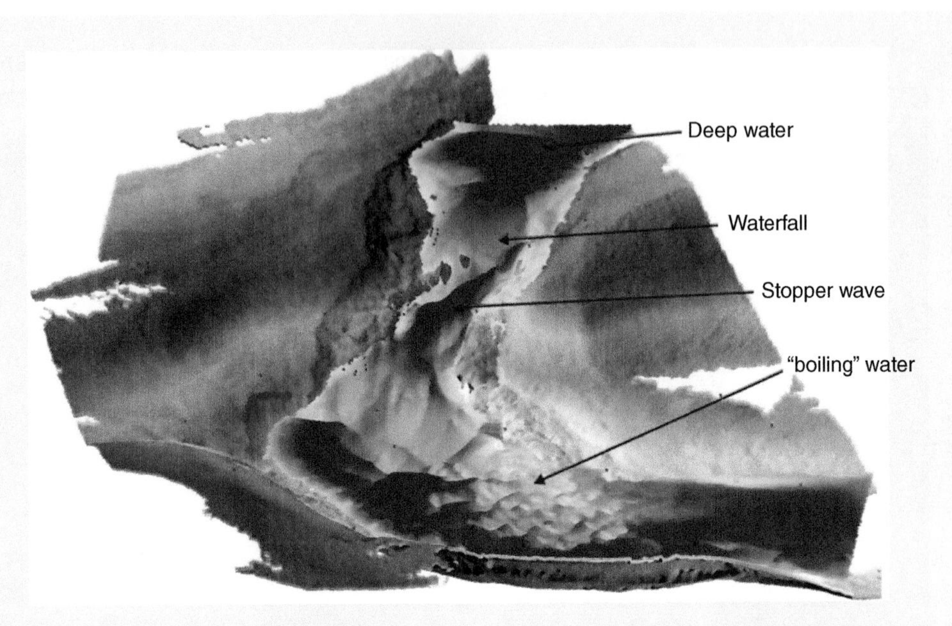


**Figure B6.3v** 3D visualisation of floodwater through a approximately 200 m long gorge, as modelled using topography gained from this project using SfM (SfM-MVS) at 0.5 m grid cell resolution. Note the capability of the model to simulate rapidly varying water surface.

dynamics (CFD) model that has been designed specifically to represent outburst flood flow characteristics. This was able to show changes in flow regime, the location of plunge pools, and areas of upwelling and recirculation in unprecedented detail for this region. The patch orthophotographs and 3D point clouds will be used to quantify bedrock erosion and deposition following future flood events, enabling the first robust assessment of geomorphological work during a jökulhlaup in this region to be made.

are that results can be obtained with far less effort than if the quantitative examination of geological cross-sections in sub-vertical cliffs comes from stereo-photogrammetry (e.g. Martín et al. 2013). In detail, the method utilising SfM-MVS-derived data uses phase congruency and phase symmetry as edge detection methods (e.g. Micklethwaite et al. 2012; Eltner et al. 2014; Vasuki et al. 2014) and user interaction to guide the process. Figure 6.2 shows semi-automated mapped fault lines overlayed onto an orthophotograph of the rock outcrop. Vasuki et al. (2014) reported that the user-guided interpretation was completed in 10 minutes, while manual digitising took approximately 7 hours. Vasuki et al. (2014) combined detected fault traces with the point cloud data to extract orientation data systematically along the faults using the RANSAC algorithm (see Section 3.4) to best-fit planes through points lying along the fault. Geological outcrop applications of SfM-MVS have also received intensive scrutiny by Gomez (2014), whose work is notable

**Figure 6.2** Fault map (green lines) resulting from semi-automated method of Vasuki et al. (2014). Image depicts a rock exposure about 20 m across. Source: From Vasuki et al. (2014).

as an example of utilising SfM-MVS-derived data to apply robust mathematical spatial statistics, wavelet analysis, to automate and objectify sediment facies (vertical sequence) description, which when measured by traditional manual methods can be tedious and often subjective.

## 6.6 Crossing Temporal Scales: Examples of Change Detection to Suggest Process Dynamics

Geoscience studies to date that have gone beyond simply creating single 3D landform models have applied SfM-MVS to detect changes in terrain, usually erosion and deposition, via repeated surveying and production of DoDs (Table 6.3). These studies tend to be targeted at rapidly moving or hazardous landforms. Several of these studies have then interpreted these changes in terrain with the consideration of other environmental data to suggest physical processes and key controlling factors. Whilst most studies emphasise spatiotemporal variability, some of the findings enabled by the application of SfM-MVS have been entirely novel; as summarised in Table 6.3.

James and Robson (2012) demonstrated a relationship between erosion rate and seasonality of erosion rates along a coastal cliff, and they found this due to obtaining data at an unprecedented spatiotemporal resolution (Table 6.3).

**Table 6.3** Selected applications of SfM-MVS in the geosciences focussed on the measurement of the changes between successive surveys.

| Environment and type of change | Notes | Measurements | Novel findings | References |
|---|---|---|---|---|
| Landslide displacement | Potentially dangerous terrain, comparison of SfM DEM with previously acquired photogrammetry-derived DEM | Horizontal surface displacements, structure from surface fissures | Fissures of different distributions and orientations relating directly to bedrock topography | Niethammer et al. (2011) |
| Landslide displacement | Extended Niethammer displacement study with surface feature tracking | COSI-corr feature tracking of horizontal surface displacements | Flow kinematics such as flow rate, landslide expansion, accumulation at toe zone, and retreating scarp | Lucieer et al. (2014) |
| Lava flow evolution | Two surveys of complex and largely inaccessible terrain | DoD and 3D coordinates of identifiable points in each survey to measure 3D displacement vectors | First observations and measurements of endogenic flow inflation, flow front advance by breakouts from insulated flow cores and transitions from juvenile slabby to mature rubbly flow textures | Tuffen et al. (2013) and James and Robson (2014b) |
| Lava dome change | Four surveys of complex and largely inaccessible terrain | DEMs and DoDs | Asymmetry of the post-explosion dome topography suggests that internal dome heterogeneity was important during the explosion | James and Varley (2012) |
| Coastal cliff erosion | Seven surveys over a year of a sub-vertical surface: point clouds converted to raster-based surfaces by transforming them into a vertically oriented cylindrical coordinate system | Semi-variogram analysis of temporally distributed erosion rates | Identification of a correlation between volume loss and time-length scale | James and Robson (2012) |
| Rill erosion | Multiple SfM DEMs constructed for detection of micro-scale spatiotemporal changes | Automatic rill extraction, rill parameter calculation, soil surface roughness, and volumetric quantification | Influence of predominant wind direction on rill development and migration, and quantification of micro-scale erosion rates | Eltner et al. (2014) |
| Gully soil erosion | TLS impractical in this setting. Three surveys spanning episodes of intense rainfall | Triangulated 3D irregular network (TIN) 3D mesh of each survey interval. Calculation of mass of soil loss between each survey | Near-complete elimination of obscuration or shadowing to derive eroded volumes by including undercuts and plunge pools into the meshed surface models | Kaiser et al. (2014) |
| Gully headcut erosion | Complex terrain, five small headcuts studied | DEMs of difference (DoDs) | Incision in main channel and spatiotemporally variable lateral bank erosion; variable erosion rates | Gomez-Gutierrez et al. (2014) |
| Periglacial stone circle surface dynamics | Micro-topography, three patches surveyed twice 3 years apart | DoDs and feature tracking of horizontal surface displacements | Difference in dynamic between inner and outer circles, and hence new conceptual model of circle spatiotemporal evolution | Kääb et al. (2013) |

Similar consideration of spatiotemporal process interrelationships enabled Eltner et al. (2014) to make novel insights into rill erosion and in particular to identify (by using edge analysis of the 3D point cloud) a control of prevailing wind direction on rill erosion for the first time (Table 6.3). Soil erosion, gully head retreat, and plunge pool development following heavy rain, and gully head retreat volume as reported by Kaiser et al. (2014), Gomez-Gutierrez et al. (2014), and Frankl et al. (2015b) all were enabled by SfM-MVS-derived DoDs and over the relative impracticality of using TLS (see Section 2.3.2) in such difficult-to-access sites (Table 6.3).

In a study of a truly inaccessible landform, James and Varley (2012) created four SfM-MVS-derived DEMs and associated DoDs for a lava dome, and analysis of the volume of the dome and mapping of linear and arcuate failure planes helped them to identify internal dome heterogeneity. Tuffen et al. (2013) took full advantage of the ultra-portability and speed of SfM-MVS techniques to produce multiple DoDs from which novel (and otherwise unobtainable) interpretation of the temporal evolution of a rhyolitic lava flow was made (see Box 6.4).

Box 6.4 Case study: Understanding the emplacement of rhyolite lava flows

Mike R. James and Hugh Tuffen, Lancaster University

### Background and context

Understanding lava flow processes is important for improving hazard management around many volcanoes. For this reason, common low-viscosity basalt lava flows have been well characterised; however, the scarcity of eruptions that produce high-viscosity rhyolite flows makes rhyolite emplacement processes difficult to study. The eruption of Cordón Caulle volcano, Chile (2011–2012) allowed the first scientific observations of an active rhyolite lava flow as it was emplaced. By capturing this emplacement in 3D, this study provided unique insight into the dynamics of a highly viscous active rhyolite flow and enabled parallels be drawn between the emplacement mechanisms observed in low-viscosity basaltic flows.

### Method

To record the evolving surface of the lava flow, approximately 600 photographs (Fig. B6.4ia,b) of the flow were taken on both January 4 and 10, 2012 using a Canon EOS 450D with a fixed 28 mm lens. Each survey was acquired by walking an approximately 1 km long return path along a ridge adjacent to the flow, allowing a good view over the active flow margin. On January 4, the survey path was recorded with a GPS track acquired with a hand-held GPS receiver to provide control data. Panoramic image sets were acquired at approximately 15 m intervals along the paths, to build up an elongate network of locally convergent imagery. The convergence was designed to minimise systematic

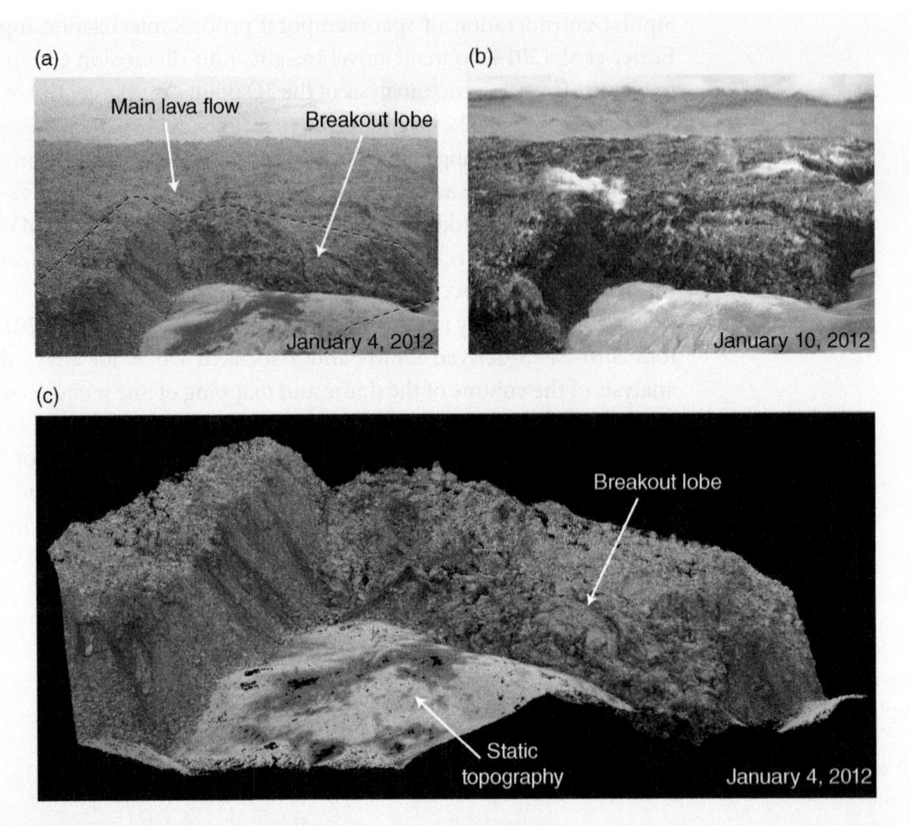


**Figure B6.4i** Examples of the images used to construct the 3D model, covering the breakout region from approximately the same position and illustrating the contrasting lighting conditions (a & b). On January 4, conditions were particularly poor due to the ongoing eruption of ash and gas. A section of the analysed point cloud (c) from the area of red-dashed outline in (a) showing the steep rubbly flow margin (left) and the smoother and thinner breakout. The width of the section shown is approximately 250 m.

error that could otherwise accumulate along the elongated projects (James & Robson 2012, 2014b). Images were processed into dense 3D point clouds (Fig. B6.4ic) using the "bundler photogrammetry package" (http://www.blog.neonascent.net/archives/bundler-photogrammetry-package/ by J. Harle) that links Bundler (Snavely et al. 2006) and PMVS2 (Furukawa & Ponce 2010). For the survey on January 10, the data were scaled and referenced with sfm_georef software (James & Robson 2012; http://www.lancs.ac.uk/staff/jamesm/software/sfm_georef.htm) by using camera positions (interpolated from the GPS track) as control data. This gave an RMSE on the camera positions of approximately 5 m, commensurate with the expected precision of the GPS receiver. Static features were identified in this scene and their coordinates obtained for use as control points on January 4, survey. Georeferencing this survey resulted in an RMSE on the control of 0.2 m.

A approximately 35 m long section of the active flow margin was selected for further analysis, and DEMs (0.5 m grid cell size) were produced from the point clouds. Vertical change was determined by differencing the DEMs, whilst image-based feature tracking (carried out in sfm_georef) was used to derive the horizontal components of motion.

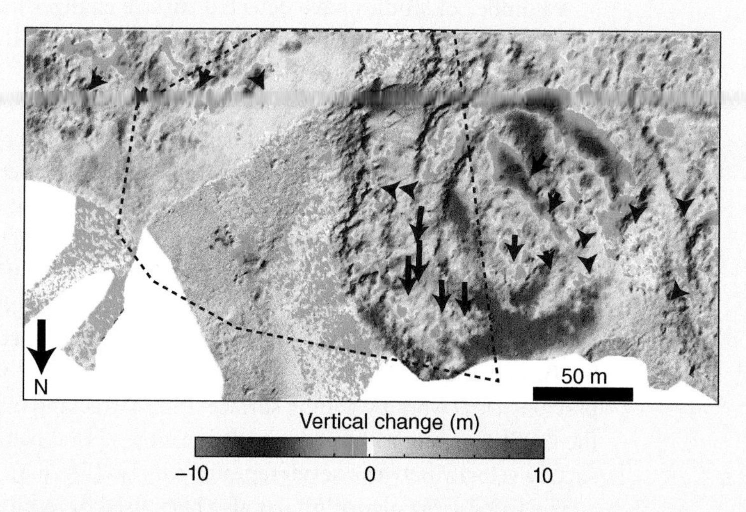


**Figure B6.4ii** Map showing vertical (colours) and horizontal (arrows) topographic change between January 4 and 10 surveys of the lava flow margin. White represents no data, and regions of less than 0.2 m vertical change are transparent to reveal the underlying hill-shaded DEM. Areas of large positive elevation increase (red) represent the advance of steeply sloping areas. Similarly, areas of significant apparent height decrease (blue) indicate horizontal motion of steep surfaces in the up-slope direction. The dashed line shows the area of the point cloud section shown in Fig. B6.4ic. Source: Adapted from Tuffen et al. (2013).

### Main findings

The DEMs produced indicated areas of flow advance and showed that a smoother region of the flow represented a highly active "breakout" area that was rapidly extruding fresh lava (Fig. B6.4ii). The main flow margin was approximately 30–40 m thick, advancing at approximately 1.5 m $day^{-1}$, and covered in a coarse rubbly surface. In contrast, the thinner breakout was characterised by a smoother surface, broken by evolving tensile fractures, and advancing at approximately 3 m $day^{-1}$. Similar breakouts are common in basaltic flows when lava continues to be erupted even after the flow front has cooled sufficiently that it stops advancing. The lava flow then evolves by thickening and widening by breakouts, changing from a "simple" flow into a "compound" flow field. This process is enabled by the increasing thermal heterogeneity of the flow; the flow surface cools, increases in viscosity, and eventually becomes solid to deform only by fracture. Meanwhile, the flow core remains hot and ductile and can breakout when weaknesses in the surrounding crust develop.

### Key points for discussion

- Ground-based photography can be used to characterise lava flow processes in 3D, with convergent imagery used to mitigate systematic error along linear tracks such as flow margins.
- Rhyolite lava flows can advance and expand by similar processes as observed in lower-viscosity basaltic flows, by the breakout and inflation of new flow lobes.
- The breakout process emphasises the influence of the cooled carapace in controlling emplacement dynamics and illustrates that the central core of the lava remains sufficiently hot to be mobile.

### Summary

The use of photo-based 3D reconstruction techniques enabled the first capture of sequential DEMs of an active rhyolite lava flow. The imaged flow processes show strong similarities with those involved during the emplacement of much lower viscosity lavas, a result that leads the way for unifying the flow emplacement processes models across all lavas.

A number of studies have detected surface changes in both the vertical and the horizontal dimensions. They have achieved this by combining feature-tracking analysis of orthoimages with analysis of DoDs. Thus whilst combining orthophotograph image analysis with point cloud analysis was discussed in Section 6.4, these studies are included here because they focus specifically on the detection and measurement of surface changes. One of the first of these "change" types of studies to demonstrate the complete SfM-MVS workflow and detailed outputs was a landslide displacement investigation by Niethammer et al. (2011). They identified areas of persistent deformation producing fissures apparently related directly to the bedrock topography. Landslide displacements were also been analysed by Lucieer et al. (2014) and by Turner et al. (2015) who working on the same site together extended previous DoD work by adding surface feature–tracking image analysis using the COSI-Corr algorithm to quantify spatiotemporal patterns in horizontal surface velocity between seven repeated surveys (Fig. 6.3).

The COSI-Corr algorithm has also been used by Whitehead et al. (2013) and by Ryan et al. (2015) to analyse SfM-MVS-derived images and DEMs to calculate mean horizontal surface velocity of a glacier, over a period of 1 year and a few weeks, respectively. Such spatially distributed velocities of a glacier surface would be otherwise extremely difficult to obtain, being restricted temporally if via standard aerial photogrammetry or satellite image remote sensing, or being restricted spatially if via ground-based survey methods. Thus even manual feature–tracking (e.g. Immerzeel et al. 2014) using successive SfM-MVS-derived orthophotographs to estimate glacier surface velocity is exceptionally useful. The same arguments of novelty, as based on spatiotemporal resolution, apply to surface displacements detected by Kääb et al.'s (2013) study on periglacial sorted circles; but unlike the glacier examples, they were able to explicitly test competing process-based hypotheses, in their case of stone circle evolution.

These "form-process" studies represent a route to understanding that is a conerstone of the geosciences. Practically, such "before" and "after" surveys will consider surfaces that are static at the time of survey and thus will employ "rigid SfM-MVS" principles. The opportunity in the geosciences for non-rigid SfM-MVS to overcome this limitation will be discussed in Section 7.8. Besides the precision of the ground control and related data processing, the interval between surveys also defines the magnitude and type of changes that can be detected. In the case of natural surfaces, if survey intervals are too short, there is a reasonable chance that no significant (above the level of uncertainty or below the minimum level of detection) change will have occurred. If survey intervals are too long, only net changes can be detected, and these may mask or hide incidences of changes that have cancelled each other out, for example erosion and subsequent infill (e.g. Carrivick et al. 2013; Vericat et al. 2014). Indeed Vericat et al. (2014) suggest that the coupling of appropriately scaled spatial and temporal data is critical to understand topographic changes, such as those resulting from sedimentological connectivity and channel network evolution, for example.

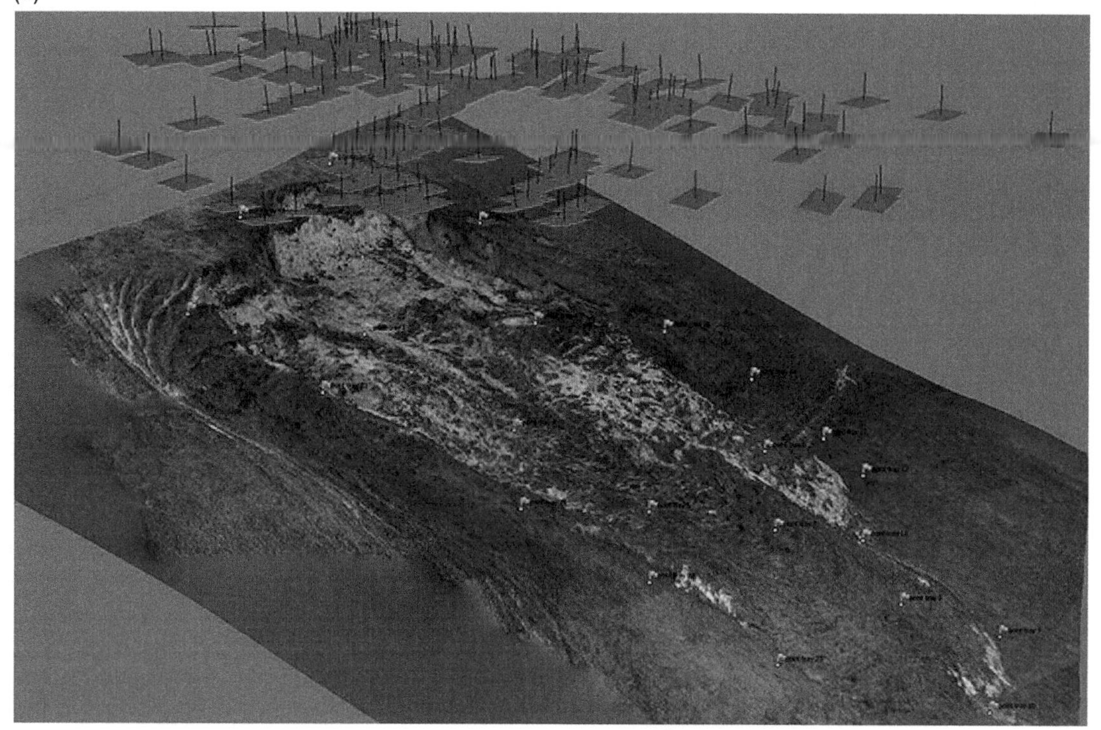


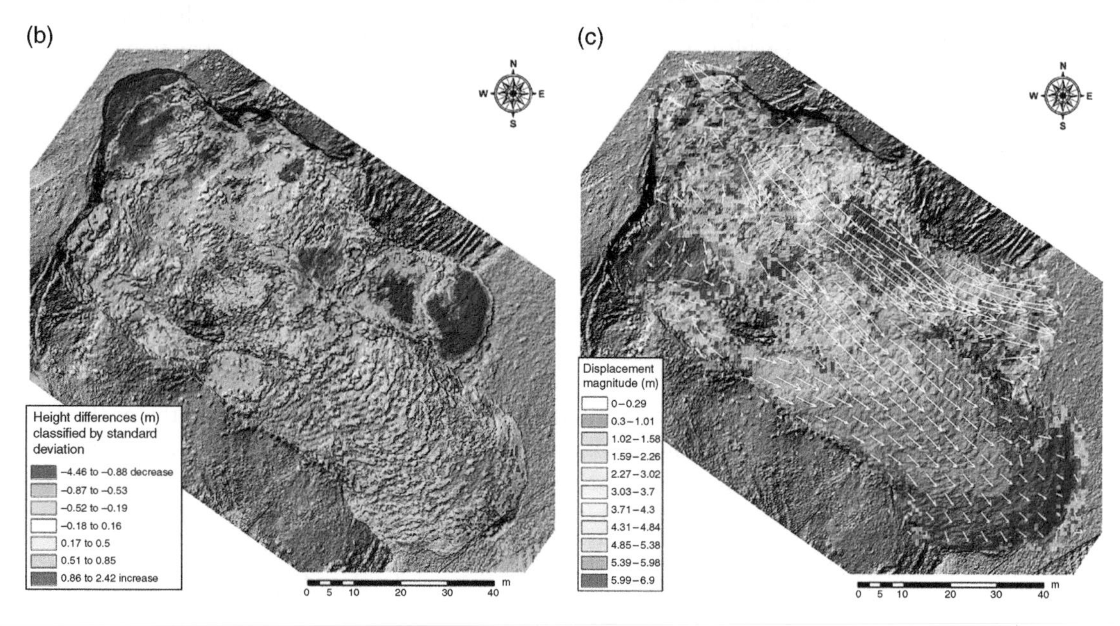


**Figure 6.3** Perspective view of a texture-mapped 3D surface (a), difference (DoD) between two digital elevation models from surveys 4 months apart (b) and calculation of the dynamics of the landslide in southeast Tasmania, especially the retreat of the main scarp and the expansion of the toes, using the statistical correlator COSI-Corr algorithm (c), by Lucieer et al. (2014). In part (a) the blue squares over the landslide show the camera positions and orientations during image acquisition by a UAV. In part (b) the numbered flags on the landslide show the positions of the ground control points used for the bundle adjustment. In part (c) the white vectors indicate displacement directions and the coloured layer illustrates the combined magnitude of displacements in the N–S and E–W directions.

Since SfM-MVS is very cheap, quick, and easy (Section 2.4), it offers the capability for high spatial resolution surveys at short survey intervals and thus high temporal resolution.

## 6.7 Practitioner-Based SfM-MVS

There are some environmental consultancies offering UAV surveys (e.g. McKinney 2015 in the United Kingdom, for example http://soarhere.com/ in the United States) and probably also SfM-MVS services, but the exact nature and especially the results of that work remain largely inaccessible, being reported only to the client. Therefore, whilst industrial applications of SfM-MVS in the geosciences certainly do exist because there are a number of commercial enterprises advertising SfM-MVS capability, most industrial work is not routinely publically reported.

Probably the first exception, which really was a test case, was that of earthwork planning by Nassar et al. (2011) who showed that ground-based image acquisition and subsequent SfM-MVS processing were most suited for pit excavations (dredging) with areas less than 2000 $m^2$ and with depths up to 5 m and for sediment piles (dumping) less than 10 m high and with base areas less than 300 $m^2$. A similar but more ambitiously scaled and automated project by Siebert and Teizer (2014) presented the application of UAV-enabled SfM-MVS to excavation and earth-moving construction sites. They noted the large scale (101 $km^2$) and the potentially hazardous nature of the site was attractive for the SfM-MVS workflow and via UAV-derived images achieved a spatial extent of 24,900 $m^2$, a total workflow (acquisition and processing) time of 165 minutes, a point cloud of greater than 2,000,000 points, and a spatial resolution/density of less than 561 points $m^{-2}$.

Perhaps the most advanced usage of SfM-MVS in the geosciences to date is that by the Jet Propulsion Laboratory (JPL 2014) Mobility and Robotic Systems Section. They have developed SfM-MVS to assist with spacecraft planetary entry, descent, and landing – a challenging time because of the extremely high and rapid reduction in spacecraft speed, little or no communications with Earth, and most importantly for the readership of this book the highly variable landing-site terrain. Failure of any system during descent and landing can result in mission failure so all systems must be extremely reliable under nominal conditions and robust to unexpected conditions.

SfM-MVS algorithms have been developed by JPL (2014) for terrain-relative navigation using passive imaging and active sensing. These systems assist in solving two fundamental entry, descent, and landing problems, namely, hazard detection and avoidance and pinpoint landing.

The JPL (2014) hazard detection and avoidance system includes dense SfM-MVS for rock and slope hazard detection (Fig. 6.4), fast detection of rocks, slopes, craters, and discontinuities with visible imagery and SfM-MVS in combination with laser scanning and phased array radar to give multi-sensor safe-site selection with fuzzy logic (JPL 2014).

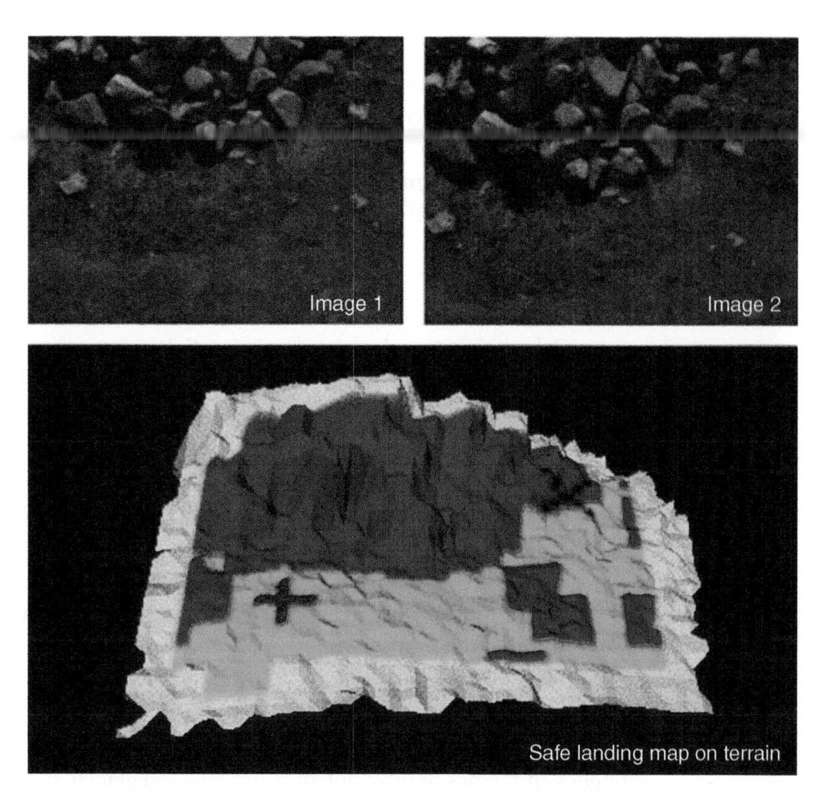


**Figure 6.4** Example of testing (of a real rock surface in a laboratory) dense Structure from Motion for rock and slope hazard detection, by the Jet Propulsion Laboratory (JPL 2014). The colours represent a classification of terrain as a possible landing site (green) and as unsuitable for landing (red). Terrain marginal to the surveyed area (yellow) is not considered to be robustly resolved for classification.

JPL (2014) report that the pinpoint landing systems utilise developments in position-estimation algorithms that match landmarks detected in data collected during descent to landmarks stored in an onboard database. With these matches, the position of the spacecraft lander can be determined relative to the surface. Algorithms based on feature tracking provide surface-relative velocity as is required to generate accurate trajectory knowledge between position measurements.

## 6.8 Summary

The geosciences are presently employing an SfM-MVS workflow to produce the following:

- Orthophotograph mosaics
- 3D point clouds
- DEMs
- DoDs

Many geoscience applications of SfM-MVS to date are essentially proof-of-concept studies aiming to produce a 3D model of a landform assumed to be static. However, some geoscience applications of SfM-MVS have realised the potential of interrogating SfM-MVS-derived data with novel algorithms for defining landform elevation, landform texture (i.e. roughness), and landform surface composition. In terms of composition, SfM-MVS-derived orthophotograph mosaics have been used for the following:

- Calculating gravel surface grain/particle sizes;
- Compositional analysis of soft sediment bedding and facies;
- Structural analysis of hard rock geology.

With the additional 3D point information, other applications of SfM-MVS that are addressing hitherto poorly constrained problems include the following:

- Definition of complex landform geometry, such as boulders and cliff or gully undercuts;
- Determining tree biomass.

SfM-MVS-derived DEMs and DoDs have been produced to detect changes and hence to infer dynamic processes in glacial, fluvial, coastal, hillslope, dryland, volcanic, and shallow underwater environments. Thereby the geosciences are realising the potential of SfM-MVS for seamlessly crossing spatial scales.

Overall, the geosciences are embracing SfM-MVS, employing it in diverse environments and in otherwise-impossible-to-access terrain to generate both orthophotograph and 3D point cloud data. Novel findings on earth surface processes are being produced, and new questions are being asked as facilitated by SfM-MVS in the geosciences. However, whilst there are now commercial enterprises offering SfM-MVS services, SfM-MVS workflows have yet to be fully embraced within industrial applications, such as for environmental consultancy.

## References

Bemis, S.P., Micklethwaite, S., Turner, D. et al. (2014) Ground-based and UAV-Based photogrammetry: a multi-scale, high-resolution mapping tool for structural geology and paleoseismology. *Journal of Structural Geology*, **69**, 163–178.

Brasington, J. (2010) From grain to floodplain: hyperscale models of braided rivers. *Journal of Hydraulic Research*, **48** (**4**), 52–53.

Brasington, J., Vericat, D. & Rychkov, I. (2012) Modeling river bed morphology, roughness, and surface sedimentology using high resolution terrestrial laser scanning. *Water Resources Research*, **48** (**11**), W11519.

Carrivick, J.L., Smith, M.W., Quincey, D.J. & Carver, S.J. (2013) Developments in budget remote sensing for the geosciences. *Geology Today*, **29** (**4**), 138–143.

Casella, E., Rovere, A., Pedroncini, A. et al. (2014) Study of wave runup using numerical models and low altitude aerial photogrammetry: a tool for coastal management. *Estuarine, Coastal and Shelf Science*, **149**, 160–167.

Castillo, C., Pérez, R., James, M.R., Quinton, J.N., Taguas, E.V. & Gómez, J.A. (2012) Comparing the accuracy of several field methods for measuring gully erosion. *Soil Science Society of America Journal*, **76** (**4**), 1319–1332.

Dandois, J.P. & Ellis, E.C. (2013) High spatial resolution three-dimensional mapping of vegetation spectral dynamics using computer vision. *Remote Sensing of Environment*, **136**, 259–276.

Eltner, A., Baumgart, P., Maas, H.G. & Faust, D. (2014) Multi-temporal UAV data for automatic measurement of rill and interrill erosion on loess soil. *Earth Surface Processes and Landforms*, **40**, 741–755.

Favalli, M., Fornaciai, A., Isola, I., Tarquini, S. & Nannipieri, L. (2012) Multiview 3D reconstruction in geosciences. *Computers & Geosciences*, **44**, 168–176.

Fonstad, M.A., Dietrich, J.T., Courville, B.C., Jensen, J.L. & Carbonneau, P.E. (2013) Topographic structure from motion: a new development in photogrammetric measurement. *Earth Surface Processes and Landforms*, **38** (**4**), 421–430.

Frankl, A., Seghers, V., Stal, C., De Maeyer, P., Petrie, G. & Nyssen, J. (2015a) Using image-based modelling (SfM-MVS) to produce a 1935 ortho-mosaic of the Ethiopian highlands. *International Journal of Digital Earth*, **8**, 421–430.

Frankl, A., Stal, C., Abraha, A. et al. (2015b) Detailed recording of gully morphology in 3D through image-based modelling. *Catena*, **127**, 92–101.

Furukawa, Y. & Ponce, J. (2010) Accurate, dense, and robust multiview stereopsis. *IEEE Transactions on Pattern Analysis and Machine Intelligence*, **32**, 1362–1376. doi: 10.1109/TPAMI.2009.161.

Gienko, G.A. & Terry, J.P. (2014) Three-dimensional modeling of coastal boulders using multi-view image measurements. *Earth Surface Processes and Landforms*, **39** (**7**), 853–864.

Gomez, C. (2014) Structure-from-Motion and Wavelet Decomposition for outcrop analysis. Technical Paper in HAL Archives en Ligne, 15 pp.

Gomez-Gutierrez, A., Schnabel, S., Berenguer-Sempere, F., Lavado-Contador, F. & Rubio-Delgado, J. (2014) Using 3D photoreconstruction methods to estimate gully headcut erosion. *Catena*, **120**, 91–101.

de Haas, T., Ventra, D., Carbonneau, P.E. & Kleinhans, M.G. (2014) Debris-flow dominance of alluvial fans masked by runoff reworking and weathering. *Geomorphology*, **217**, 165–181.

Harwin, S. & Lucieer, A. (2012) Assessing the accuracy of georeferenced point clouds produced via multi-view stereopsis from unmanned aerial vehicle (UAV) imagery. *Remote Sensing*, **4** (**6**), 1573–1599.

Hugenholtz, C.H., Whitehead, K., Brown, O.W. et al. (2013) Geomorphological mapping with a small unmanned aircraft system (sUAS): feature detection and accuracy assessment of a photogrammetrically-derived digital terrain model. *Geomorphology*, **194**, 16–24.

Hutchinson, M. & Gallant, J.C. (2000) Digital elevation models and representation of terrain shape. In: J.P. Wilson & J.C. Gallant (eds), *Terrain Analysis: Principles and Applications*, pp. 29–50. John Wiley & Sons, Inc., New York.

Immerzeel, W.W., Kraaijenbrink, P.D.A., Shea, J.M. et al. (2014) High-resolution monitoring of Himalayan glacier dynamics using unmanned aerial vehicles. *Remote Sensing of Environment*, **150**, 93–103.

James, M.R. & Robson, S. (2012) Straightforward reconstruction of 3D surfaces and topography with a camera: accuracy and geoscience application. *Journal of Geophysical Research, Earth Surface*, **117**, F03017.

James, M.R. & Robson, S. (2014) Sequential digital elevation models of active lava flows from ground-based stereo time-lapse imagery. *ISPRS Journal of Photogrammetry and Remote Sensing*, **97**, 160–170. doi: 10.1016/j.isprsjprs.2014.08.011.

James, M.R. & Varley, N. (2012) Identification of structural controls in an active lava dome with high resolution DEMs: Volcán de Colima, Mexico. *Geophysical Research Letters*, **39** (**22**), L22303.

Javernick, L., Brasington, J. & Caruso, B. (2014) Modelling the topography of shallow braided rivers using Structure-from-Motion photogrammetry. *Geomorphology*, **213**, 166–182.

Johnson, K., Nissen, E., Saripalli, S. et al. (2014) Rapid mapping of ultrafine fault zone topography with structure from motion. *Geosphere*, **10**, 969–986.

JPL Robotics (2014) *Application: landing*. http://www-robotics.jpl.nasa.gov/applications/applicationArea.cfm?App=3 [accessed on November 14, 2014].

Kääb, A., Girod, L. & Berthling, I. (2013) Surface kinematics of periglacial sorted circles using Structure-from-Motion technology. *The Cryosphere Discussions*, **7** (**6**), 6043–6074.

Kaiser, A., Neugirg, F., Rock, G. et al. (2014) Small-scale surface reconstruction and volume calculation of soil erosion in complex Moroccan gully morphology using structure from motion. *Remote Sensing*, **6** (**8**), 7050–7080.

Leon, J.X., Roelfsema, C.M., Saunders, M.I. & Phinn, S.R. (2015) Measuring coral reef terrain roughness using 'Structure-from-Motion' close-range photogrammetry. *Geomorphology*, **242**, 21–28.

Lisein, J., Linchant, J., Lejeune, P., Bouché, P. & Vermeulen, C. (2013) Aerial surveys using an Unmanned Aerial System (UAS): comparison of different methods for estimating the surface area of sampling strips. *Tropical Conservation Science*, **6** (**4**), 506–520.

Lucieer, A., de Jong, S. & Turner, D. (2014) Mapping landslide displacements using Structure from Motion (SfM-MVS) and image correlation of multi-temporal UAV photography. *Progress in Physical Geography*, **38**, 97–116.

Mancini, F., Dubbini, M., Gattelli, M., Stecchi, F., Fabbri, S. & Gabbianelli, G. (2013) Using unmanned aerial vehicles (UAV) for high-resolution reconstruction of topography: the structure from motion approach on coastal environments. *Remote Sensing*, **5** (**12**), 6880–6898.

Martín, S., Uzkeda, H., Poblet, J., Bulnes, M. & Rubio, R. (2013) Construction of accurate geological cross-sections along trenches, cliffs and mountain slopes using photogrammetry. *Computers & Geosciences*, **51**, 90–100.

McKinney, S. (2015) The flying eye. In: E. Yarrow (ed), *The Environment*, **20**, pp. 34–35. CIWEM, London.

Meesuk, V., Vojinovic, Z., Mynett, A.E. & Abdullah, A.F. (2015) Urban flood modelling combining top-view LiDAR data with ground-view SfM-MVS observations. *Advances in Water Resources*, **75**, 105–117.

Micklethwaite, S., Turner, D., Vasuki, Y., Kovesi, P., Holden, E.-J. & Lucieer, A. (2012) Mapping from an Armchair: rapid, high-resolution mapping using UAV and computer vision technology. *Structural Geology & Resources*, 130–133.

Nassar, K., Aly, E. A. & Jung, Y. (2011) Structure-from-motion for earthwork planning. In: *Proceedings of the 28th ISARC*, June 29–July 2, 2011, pp. 310–316. Seoul, Korea.

Nicosevici, T. & Garcia, R. (2008) Online robust 3D mapping using structure from motion cues. In: *OCEANS 2008-MTS/IEEE Kobe Techno-Ocean*, April 2008, pp. 1–7. IEEE.

Niethammer, U., Rothmund, S., Schwaderer, U., Zeman, J. & Joswig, M. (2011) Open source image-processing tools for low-cost UAV-based landslide investigations. *International Archives of the Photogrammetry, Remote Sensing and Spatial Information Sciences*, **38** (**1**), C22.

Niethammer, U., James, M.R., Rothmund, S., Travelletti, J. & Joswig, M. (2012) UAV-based remote sensing of the Super-Sauze landslide: evaluation and results. *Engineering Geology*, **128**, 2–11.

Rippin, D.M., Pomfret, A. & King, N. (2015) High resolution mapping of supra-glacial drainage pathways reveals link between micro-channel drainage density, surface roughness and surface reflectance. *Earth Surface Processes and Landforms*, **40** (**10**), 1279–1290.

Ružić, I., Marović, I., Benac, Č. & Ilić, S. (2014) Coastal cliff geometry derived from structure-from-motion photogrammetry at Stara Baška, Krk Island, Croatia. *Geo-Marine Letters*, **34** (**6**), 555–565.

Ryan, J.C., Hubbard, A.L., Box, J.E. et al. (2015) UAV photogrammetry and structure from motion to assess calving dynamics at Store Glacier, a large outlet draining the Greenland ice sheet. *The Cryosphere*, **9** (**1**), 1–11.

Rychkov, I., Brasington, J. & Vericat, D. (2012) Computational and methodological aspects of terrestrial surface analysis based on point clouds. *Computers & Geosciences*, **42**, 64–70.

Siebert, S. & Teizer, J. (2014) Mobile 3D mapping for surveying earthwork projects using an Unmanned Aerial Vehicle (UAV) system. *Automation in Construction*, **41**, 1–14.

Smith, M.W., Carrivick, J.L., Hooke, J. & Kirkby, M.J. (2014) Reconstructing flash flood magnitudes using "structure-from-motion": a rapid assessment tool. *Journal of Hydrology*, **519**, 1914–1927.

Snavely, N., Seitz, S. M., & Szeliski, R. (2006, July). Photo tourism: exploring photo collections in 3D. In: *ACM Transactions on Graphics* (*TOG*), Vol. **25**, No. 3, pp. 835–846. ACM.

Stumpf, A., Malet, J.P., Allemand, P., Pierrot-Deseilligny, M. & Skupinski, G. (2015) Ground-based multi-view photogrammetry for the monitoring of landslide deformation and erosion. *Geomorphology*, **231**, 130–145.

Tuffen, H., James, M.R., Castro, J.M. & Schipper, C.I. (2013) Exceptional mobility of a rhyolitic obsidian flow: observations from Cordón Caulle, Chile, 2011-2013. *Nature Communications*, **4**, 2709. doi: 10.1038/ncomms3709.

Turner, D., Lucieer, A. & de Jong, S.M. (2015) Time series analysis of landslide dynamics using an unmanned aerial vehicle (UAV). *Remote Sensing*, **7** (**2**), 1736–1757.

Vasuki, Y., Holden, E.J., Kovesi, P. & Micklethwaite, S. (2014) Semi-automatic mapping of geological Structures using UAV-based photogrammetric data: an image analysis approach. *Computers & Geosciences*, **69**, 22–32.

Vericat, D., Smith, M.W. & Brasington, J. (2014) Patterns of topographic change in sub-humid badlands determined by high resolution multi-temporal topographic surveys. *Catena*, **120**, 164–176.

Westoby, M.J., Brasington, J., Glasser, N.F., Hambrey, M.J. & Reynolds, J.M. (2012) 'Structure-from-motion' photogrammetry: a low-cost, effective tool for geoscience applications. *Geomorphology*, **179**, 300–314.

Whitehead, K. & Hugenholtz, C.H. (2014) Remote sensing of the environment with small unmanned aircraft systems (UASs), part 1: a review of progress and challenges 1. *Journal of Unmanned Vehicle Systems*, **2** (**3**), 69–85.

Whitehead, K., Moorman, B.J. & Hugenholtz, C.H. (2013) Brief communication: low-cost, on-demand aerial photogrammetry for glaciological measurement. *The Cryosphere*, **7** (**6**), 1879–1884.

Williams, R.D., Brasington, J., Vericat, D. & Hicks, D.M. (2014) Hyperscale terrain modelling of braided rivers: fusing mobile terrestrial laser scanning and optical bathymetric mapping. *Earth Surface Processes and Landforms*, **39** (**2**), 167–183.

Woodget, A.S., Carbonneau, P., Visser, F. & Maddock, I. (2015) Quantifying submerged fluvial topography using hyperspatial resolution UAS imagery and structure from motion photogrammetry. *Earth Surface Processes and Landforms*, **40**, 47–64.

## Further Reading/Resources

Micheletti, N., Chandler, J.H. & Lane, S.N. (2015) Section 2.2.2: structure for motion. In: L.E. Clarke & J.M. Nield (eds), *Geomorphological Techniques (Online Edition)*. British Society for Geomorphology, London.

# Developing Structure from Motion in the Geosciences

## *Future Directions*


### Abstract

Structure from Motion - Multi-View Stereo (SfM-MVS) is currently a nascent technology in application to the geosciences: there is clear potential for further development. Inevitably hardware will become cheaper and lighter in the future, computer speeds will increase, and cameras will contain more advanced sensors. However, allied to these developments, the geosciences can rapidly advance its usage of SfM-MVS by adapting ongoing developments in other disciplines. Progressive automation of acquisition of images will in the future be enabled by the use of multiple time-synchronised cameras simultaneously, by automated video capture, by cameras triggered by an environmental sensor, and by crowd-sourced images. As data capture potential expands, image organisation will become increasingly important. Improved segmentation of images and improved recognition of objects or features within images could be utilised by the geosciences not only to improve SfM-MVS workflow efficiency but also to improve automatic classification and analysis of surfaces. The potential for the geosciences to utilise real-time maps and real-time three-dimensional (3D) surface models is vast; feature detection in the field, surface classification in the field, and quantification of differences in surface elevation in the field, for example, would all make interpretations of environmental processes much more objective and hence would make understanding more immediate and more complete. The geosciences should explore augmented reality environments as an alternative to proprietary point cloud software and geographical information system (GIS) software for 3D point cloud manipulation and visualisation. Arguably non-rigid SfM-MVS is the biggest as-yet unrealised and most exciting development related to SfM-MVS that could occur for the geosciences.



### Keywords

Structure from Motion; instantaneous SfM; non-rigid SfM; SLAM; crowd-sourcing; point cloud; augmented reality

*Structure from Motion in the Geosciences*, First Edition. Jonathan L. Carrivick, Mark W. Smith, and Duncan J. Quincey.



Companion Website: www.wiley.com/go/carrivick/structuremotiongeosciences

## 7.1 Introduction

In order to look into the future, in this chapter we cast our observations beyond the geosciences to speculate on what currently emerging methods and technologies that are allied to Structure from Motion - Multi-View Stereo (SfM-MVS) could be borrowed, adapted, and developed for use in the geosciences. As historically with most technology, the military has been at the forefront of application-driven SfM-MVS technological development. By way of example, the SfM-MVS technique has been exploited by the military with vision sensors mounted on a moving robotic vehicle that computed three-dimensional (3D) geometry from observed two-dimensional (2D) features over several frames or views, and visual cues were provided to a "situation awareness" system for further tracking and recognition of moving objects (Shim et al. 2008). That SfM-MVS framework was capable of providing robust perception functions, such as ranging for autonomous mobility, mid-range sensing for tactical behaviour, moving target indication, and appearance-based automatic target recognition. Each of these functions, and in combination, could have applications in the geosciences.

Elsewhere, SfM-MVS and similar technologies are moving out of the field of computer vision and becoming applied in more varied situations. Examples are too numerous to list comprehensively or reference here but include sports visualisation and analysis, animation, anatomical surveys, documentation and digital preservation of archaeological sites and artefacts, engineering planning and analysis, and engineering robotic applications. The reasons why uptake of SfM-MVS is rapidly becoming embraced by the geosciences have been summarised in Chapter 2. This chapter is forward looking but is deliberately avoiding incremental developments. Rather, it is more ambitious, firstly looking at developments in SfM-MVS in other disciplines and then secondly using this information to suggest specific applications where geoscientists may make step changes in knowledge. In particular, it is emphasised that geosciences should consider applications of SfM-MVS beyond a static or rigid 3D reconstruction of a landform. A static reconstruction of landform may not be the final goal of a project, but rather an important intermediate step for process-based analyses of change detection, tracking, pattern and texture recognition, and modelling.

In detail, this chapter firstly describes some developments in hardware (Section 7.2), acquisition of images (Section 7.3), and software and processing methodology (Sections 7.4 and 7.5) that can be expected or is required for geoscience usage. Secondly, in Sections 7.6 through to 7.7, this chapter discusses major project types that have yet to be exploited fully by the geosciences, namely, automatic detection, real-time mapping, augmented reality, remote or inaccessible surveying, continuous monitoring, surveys of moving surfaces, and combining SfM-MVS products with other remotely sensed data.

## 7.2 Developments in Hardware

### 7.2.1 *Platforms*

It could certainly be argued that the SfM-MVS surveying revolution in the geosciences has been enabled by unmanned aerial vehicles (UAVs). A lot of progress in small (<25 kg) UAV platforms has been made in a very short space of time since Hardin and Hardin (2010, p. 1297) considered that "significant limitations of small-scale [UAVs] include their low stability as photographic platforms, short flight times, airframe fragility, the paucity of sensor packages available, and difficulties involved integrating pilot-assist flight navigation systems." Multi-rotor UAVs now exist across scales of size (see Section 4.2.2 and Fig. 4.2) from small "palm-of-the-hand" toys to "full-size" systems such as the e-volocopter (http://www.e-volo.com/) that is capable of transporting payloads equivalent to an adult person. Therefore, the issues raised in the quote could be considered to be largely resolved as far as SfM-MVS applications and use of UAVs are concerned.

Automated take-off and landing technology for UAVs is improving, but is not foolproof, and significant pilot skill may be required during those critical flight phases. In contrast, and as Hardin and Jensen (2011) noted, the technology to maintain a small vehicle in straight and level flight at a predefined altitude is mature, and these systems significantly reduce pilot burden. However, Hardin and Jensen (2011) also note that these avionics are not yet adequate to prevent a lot of image "noise" of even image loss due from wind, turbulence, and vibration through the UAV airframe; this is a broad avenue of opportunity for continued research and development. GPS-enabled and automated guidance systems for aircraft navigation through a set of predefined waypoints are also available. The precision with which these predefined routes can be flown using consumer-grade GPS receivers is not sufficient for higher-precision requirements, such as low-altitude photography at predefined coordinate locations along a route; obtaining the desired flight path precision remains a challenge with current technology. Nonetheless, consumer-grade GPS receivers do help to achieve consistency in multi-temporal surveys.

For further reading on UAV development, the reviews by Whitehead and Hugenholtz (2014) and Colomina and Molina (2014) identify limitations of the current generation of platforms and sensors, some key research challenges, and the necessary development of optimal methodologies for processing and analysis, some of which we discuss further in Sections 7.4 and 7.5. We draw attention to Box 5.1 that details the use of manned gyrocopters as changing legislation in the United Kingdom allows them to be used from April 2015. We consider that whilst ground-based remotely operated vehicles (ROVs) are unlikely to see widespread uptake for SfM-MVS in the geosciences because of their limited (low-angle) field of view, they might find utility for specifically difficult-to-access sites such as the base of cliffs, for example. In contrast, underwater ROVs are up-and-coming in technological development, affordability, and in use as SfM-MVS platforms.

### *7.2.2 Cameras: Stills versus Video*

Most SfM-MVS applications to date have used consumer-grade digital cameras. This is not surprising because consumer-grade digital cameras are small, lightweight, require less power, and have the potential to store hundreds of images, making them ideal for usage onboard small UAVs or when hand held. Still cameras presently have a resolution that is already far greater than apparently necessary for SfM-MVS, especially at close range, and whilst at present most images have to be "degraded" before use in SfM-MVS software to cope with processing power that may not be necessary in the future. However, a digital single-lens reflex (DSLR) camera sensor is small when compared with old film-based cameras, and so it is worth considering that new "fx" sensors have a "deeper" perspective/higher sensitivity because they compress the light less. The use of survey-grade aerial photography metric cameras that hardly compress the light at all would seem to be advantageous for SfM-MVS in the geosciences, especially when taking images from large survey ranges. Additionally, the geosciences might in future consider potential applications that could benefit from developments in 360° lens cameras, such as https://www.panono.com/, and from developments in corresponding software for viewing such 360° images.

In contrast, although the use of video from small UAV platforms is widespread (e.g. Jensen et al. 2008), the use of aerial video with respect to SfM-MVS in the geosciences has been speculated (Fonstad et al. 2013; Chapter 4) but remains unreported. Real-time transmission of video from a UAV to a ground station, laptop screen, dedicated monitor, or goggles is particularly attractive for SfM-MVS topographic surveying for acquisition, management, and manipulation of video imagery. It might be possible to envisage a future where goggles displayed an SfM-MVS landform reconstruction on-the-fly, that is, in real time, to allow for targeted image acquisition.

In terms of acquisition of video, Hardin and Jensen (2011) summarise the advantages that it can (i) be used for coarse-resolution data collection, (ii) facilitate visual aircraft navigation over the target, and (iii) transmit information about the UAV status, such as geographical position and battery level. Management and manipulation of imagery is discussed in Section 7.4, but in terms of video, specifically, it is perhaps useful to note here that there are promising open-source software developments for automated mosaicking video, reconstructing video, stabilising video, undistorting video, and tracking points in video. For further information, a useful overview of video-based 3D scene reconstruction, including SfM-MVS methods, is incorporated within the review by Smolic et al. (2011) on 3D video post-production and processing.

The untapped opportunity of the use of video for SfM-MVS in the geosciences can be illustrated with the widespread use of video in urban settings for generating texture-mapped 3D city models (e.g., see the review by Musialski et al. (2013)). This information is automatically extracted from the imagery collected by survey vehicles, which are equipped with cameras, GPS units, and odometry sensors, and drive around daily to record new city data. These urban car-mounted systems could reasonably be adapted for simultaneous imaging of narrow valley sides, gorges, canyon walls, or opposing river banks, for example. The urban models themselves will be exceptionally useful for micro-scale analyses of urban flooding problems.

### 7.2.3 Positioning

Outdoor global positioning with a differential Global Positioning System (dGPS) or local positioning by total station (TS) (Chapter 2) to obtain ground control points (GCPs) will continue to improve in speed and accuracy as these technologies develop incrementally.

Indoors, the geosciences has applications for SfM-MVS in experiments, such as in flumes for examining sediment transport. 3D spatial surveys must presently negotiate the limitations of optical survey instruments that when mounted on tripods are cumbersome and are limited to line of sight, which is often restricted. However, giant leaps are presently being made with Bluetooth-enabled positioning technology, most notably with iBeacons (2015), which are low-energy devices emitting identification signals at as low as 1 Hz or as fast as every 100 ms and following a strict format to give a categorised range, namely, immediate: within a few centimetres, near: within a couple of metres, and far: greater than 10 m away. The extremely low power consumption of iBeacons makes them very attractive and is not unimaginable that the technology will develop to give more precise (absolute) distance measurements in the future.

## 7.3 Progressive Automation of Acquisition

Aside from the automation in UAV platform navigation as mentioned in Section 7.2, there is also considerable scope for geoscience applications to benefit greatly from development in automation of image acquisition. The benefits would be in terms of efficiency by both reducing personnel time in the field and by removing the tricky logistics involved with gaining pre- and post-event surveys for rapidly changing parts of the earth surface. A future field system could perform SfM-MVS on the fly and trigger an alarm when significant change was detected or expected in a quarry, for landslide or avalanche monitoring, or for glacier calving, for example. Better constraining the timing of events would result by substantially reducing the survey interval. Automation could be in terms of developing the use of multiple cameras simultaneously, externally triggered cameras, crowd-sourced images, and autonomous UAV flights, for example.

### 7.3.1 *Use of Multiple Cameras Simultaneously*

Some geophysical phenomena are too inaccessible, too far from habitation, too slow, or even too fast to survey or to consider re-survey. Thus in order to remove the requirement for personnel to physically be on site conducting a survey, either with a hand-held camera or with a camera mounted on an aerial platform, there is the option to deploy multiple cameras in a static array. Multiple camera arrays are relatively incremental in advancing the use of SfM-MVS in the geosciences but is included in this chapter because no published examples yet exist of this survey method.

A static array of cameras needs careful designing which can be facilitated by the use of geographical information system (GIS) (Fig. 7.1). Pre-survey planning is worthwhile to provide certainty that suitable SfM-MVS data will be produced from the static array, as discussed in Box 7.1. A provisional pre-fieldwork GIS-based workflow to design a static camera array would include (i) gaining a coarse digital elevation model (DEM) of the area of interest, (ii) conducting a buffer analysis of the central point of interest to determine survey baseline distances, (iii) conducting a view shed analysis from that central point of interest to determine suitable camera positions, (iv) noting the field of view of the make/model of camera to be used to check for sufficient overlap in images from adjacent camera positions, and (v) estimating the image pixel size at the specific object of interest. The nature of the phenomena to be studied also needs consideration with the likely changes in surface position determining the best geometry of the cameras, for example side looking to capture horizontal motion.

Key hardware requirements for cameras in a static array are (i) to be capable of being powered for long periods of time, (ii) to have a time-lapse facility permitting single-shot images to be programmed to be taken at specified intervals/times, (iii) to have suitable memory space for the images until a physical download is possible, and (iv) to be rugged and preferably weatherproof. It is not hard to imagine a UAV capable of hibernating before flying a predefined path at a determined survey interval (legislation limitations aside). Static arrays can also be used to capture exceptionally fast-moving surfaces – for example, a water surface and a lava flow experiment as presented by Dietrich in Box 7.2 – but camera synchronisation, camera

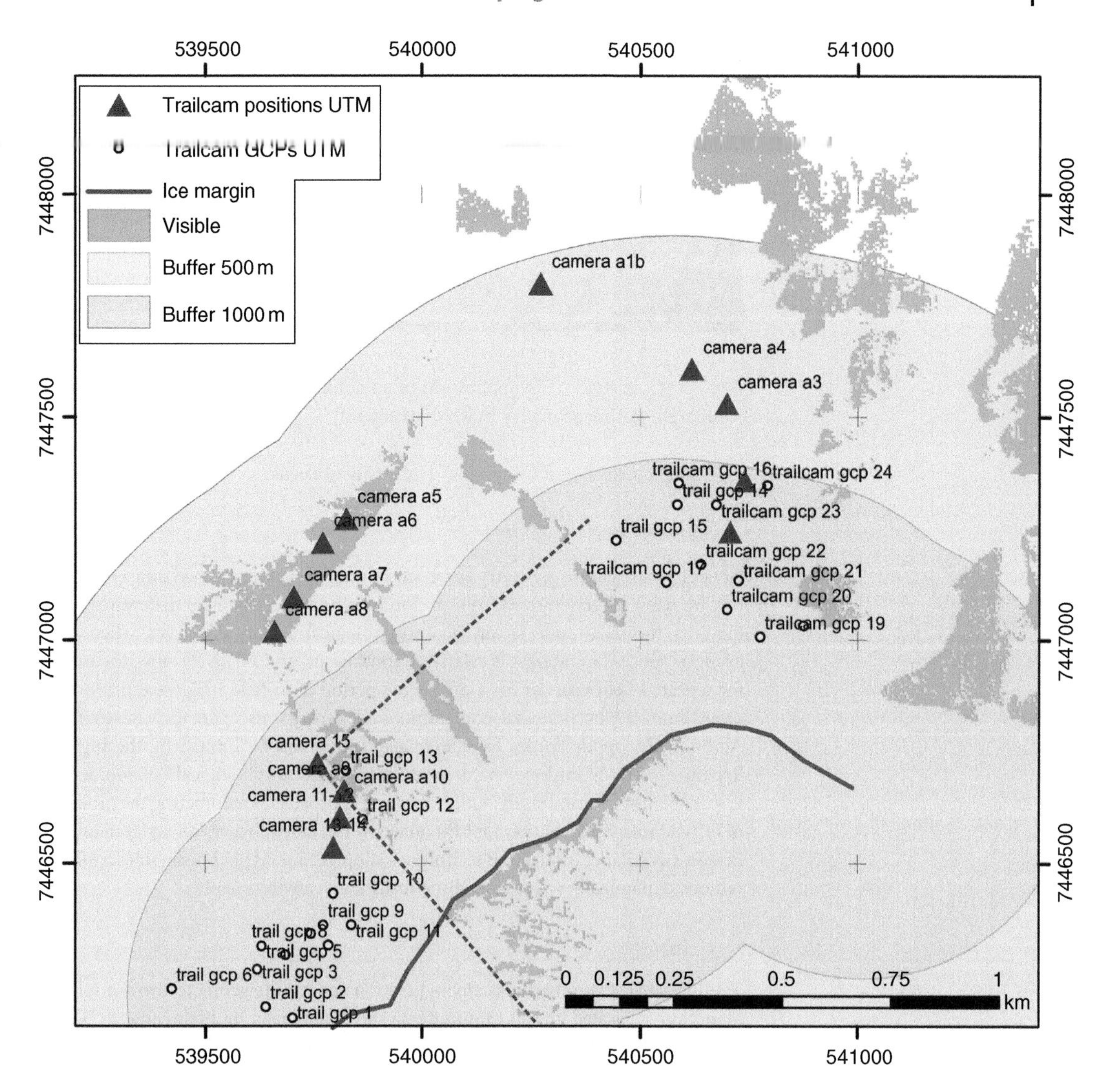


**Figure 7.1** Survey layout design for an array of trail cameras, one of which is depicted in Fig. 4.5, to image a glacier edge at an ice-marginal lake at Russell Glacier, western Greenland. Note only one viewshed and one 90 degree angle analysis are depicted for clarity.

clock drift, and processing power are challenges that need to be overcome for such applications.

Rather than all cameras within an array taking images at predefined intervals or at predefined times, it is possible that image capture from such arrays could be "triggered" via an environmental sensor. For example, a rain gauge if interested in rain splash erosion, a river flow level if interested in river bank erosion, and an anemometer if interested in sand dune migration. Alternatively and with some thought given to changing visibility due to light and weather conditions, a single camera could be the sensor whereby if a significant

Box 7.1 Case study: The application of an automated camera array to generate ice-margin feature geometry in west Greenland

Joe Mallalieu, School of Geography, University of Leeds

### Background and context

Generation of feature geometry from sets of images captured at two time episodes can be used to generate 3D point clouds, which can then be differenced to quantify the magnitude of any geomorphological activity. However, when the event or process of interest is relatively continuous, and relatively slow, the need for repeated field visits or for a prolonged period of in-field image acquisition is often hindered by financial constraints and perhaps also climatic constraints. Additionally, opportunities for image acquisition can be limited by the highly dynamic and hazardous nature of many geomorphological phenomena. Consequently, this case study outlines a novel approach for automating the process of in-field image acquisition for SfM analysis. The study focuses on an ice-margin in west Greenland, though the techniques adopted should be transferable to other remote, dynamically active, and climatically harsh environments.

### Method

Traditional SfM reconstructs camera position and feature geometry from a series of motion-separated images captured by a single camera or multiple cameras. This case study illustrates an alternative approach to simulate the motion of a surface of interest between successive image sets by establishing a "fixed camera array." This array consists of a series of cameras each oriented towards the surface of interest with sufficient image overlap to permit topographic reconstruction. The frequency, timing, and duration of image acquisition can all be dictated via an automation of the array.

The fixed camera array was developed to facilitate the analysis of seasonal changes in ice-margin topography for a lacustrine-terminating section of the ice sheet margin in western Greenland. The array was equipped with 15 LtL Acorn 5210 trail cameras stationed 100–1000 m from the ice margin in a broad arc around the lake shore (Fig. B7.1i). The camera model was chosen for its high megapixel count (12 MP), fully programmable timer, and relative affordability (~£120). Each camera was powered internally by eight lithium AA batteries, though an external battery source can be connected for longer periods of image acquisition. The cameras were installed at the beginning of the melt season and programmed to capture images at 08:00, 12:00, and 16:00 hours daily. All camera

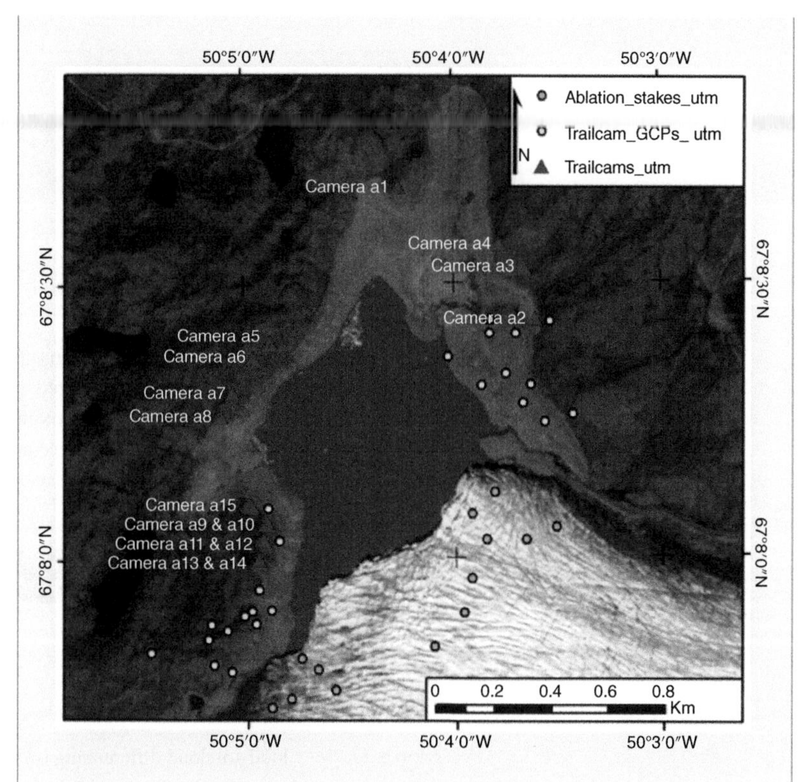


**Figure B7.1i** Overview of the ice-dammed lake and associated ice margin, and camera array setup, including camera and GCP positions.

positions and a further 24 GCPs (Fig. B7.1i) within the field of view of at least three cameras were surveyed using differential GPS to sub-centimetre accuracy.

The camera imagery (an example of which is the banner image to this case study) was downloaded manually and processed in Agisoft PhotoScan Professional to reconstruct point clouds of ice-margin geometry. The point clouds were georeferenced using the GPS measurements of the GCPs. Point clouds were subsequently differenced in CloudCompare to reveal changes in ice-margin topography over a range of timescales.

## Main findings

The return visit to the field site 10 months after installation found that 14 of the 15 cameras were still fully operative having survived repeated burial by snow and temperatures as low as −33°C. The imagery downloaded from the cameras revealed a continuous record of ice-margin dynamics spanning 301 days. Image processing in Agisoft PhotoScan Professional subsequently permitted extensive construction of ice-margin feature geometry – including the rendering of small topographic features such as terraces and undercuts – from only 14 images. The resultant dense point clouds achieved a spatial resolution of approximately 25 points per square metre on the ice-face (Fig. B7.1ii). Upon georeferencing total absolute error in the point clouds was approximately 1 m.

**Figure B7.1ii** 3D perspective view of a dense point cloud of the ice-margin geometry generated from 15 images captured at midday on August 7, 2014. This ice-margin 3D model comprises approximately $1.8 \times 10^6$ points. An interactive example of a point cloud of an ice margin in west Greenland is available on the companion website, courtesy of Joseph Mallalieu.

**Figure B7.1iii** Cloud-to-cloud differencing (in metres) derived from images captured at midday on the August 4 and 7, 2014. The black circle highlights the small calving event in illustrated Figure B7.1iv.

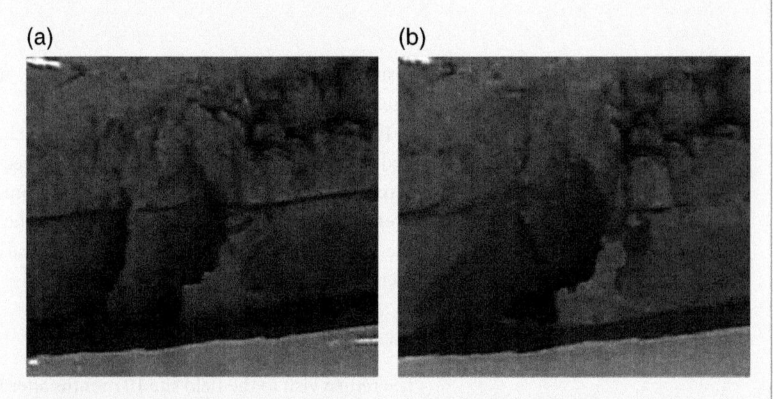


**Figure B7.1iv** Small calving event occurring between August 4 and 7, 2014 (a and b, respectively).

Tentative cloud-to-cloud differencing of ice-margin geometries in CloudCompare has shown the potential of the technique to quantify changes in ice-margin position across a range of timescales and document the magnitude and frequency of calving events (Figs. B7.2iii & B7.2iv). Due to the relatively low number of images run through the SfM software, the total processing time to generate and differentiate two point clouds can be under 2 hours. Validation of this technique for quantifying geometric change is currently ongoing via comparison with a record of ice-margin dynamics generated from dGPS surveys of ice-surface stakes shortly after camera installation.

### Key points for discussion

- The application of a fixed camera array permits the detection and measurement of incremental changes in geomorphological phenomena over a prolonged period without recourse to sustained or frequent field visits.
- The derivation of feature geometry from such a fixed camera array is far faster than alternative remote-sensing techniques (e.g. terrestrial laser scanning). Furthermore, it is capable of providing finer spatial resolution than alternative satellite or airborne sources (e.g. airborne laser scanning) and of providing fully 3D information of undercuts and concavities in sub-vertical cliff faces.
- The application of fixed camera arrays is likely to become more attractive as higher specification cameras and options for wireless image transmission become increasingly affordable.

### Summary

The application of an automated fixed camera array to generate ice-margin feature geometry in western Greenland has highlighted the setup as a valuable alternative to existing remote sensing techniques and traditional in-field image acquisition, particularly where fieldwork is limited by time, financial, or climatic constraints.

## Box 7.2 Case study: Instantaneous Structure from Motion for dynamic geomorphology

James T. Dietrich, William H. Neukom Institute for Computational Science and Department of Geography, Dartmouth College

### Background and context

One of the fundamental limitations of collecting SfM imagery with a single camera is that if any part of the scene is not stationary, the resulting point cloud often has very noisy or incomplete areas. For static scenes, the moving elements may not be important, but for scenes where the moving elements are the object of study, the results are usually not satisfactory. One solution to this problem is to use multiple simultaneously triggered cameras to capture multiple instantaneous views of the subject, which can generate an instantaneous SfM (ISfM) reconstruction. The real power of ISfM is that it produces 3D time-lapse data sets that provide not only the structure of moving subjects but also important information in the rates of change. Some of the potential applications include water surfaces (elevation and slopes) (Chandler et al. 2008), lava flows (James & Robson 2014b), mass movements (landslides/debris flows) (Bitelli et al. 2004), vegetation structure, sediment transport, laboratory and physical model experiments, and longer-term time-lapse monitoring.

### Method

ISfM holds to the same standards of one-camera SfM, mainly that the quantity and quality of the imagery and ground control determine the ultimate quality of the 3D data. For ISfM there is no real upper limit to the number of cameras used in an ISfM array. From our experiments and those of James and Robson (2014b), the lower limit is two cameras, with a manageable number of cameras ranging from 6 to 20 depending on the size of the scene and the required resolution. Additional cameras are able to provide a wider range of views that help reduce shadowing and increase the accuracy and resolution of the reconstruction. The choice of camera type, point and shoot versus DSLR, has several implications for the design and function of the array. Point-and-shoot cameras are inexpensive, compact, and lightweight – all features important for mobile deployments. The main disadvantage of using point-and-shoot cameras is that they often need modifications for synchronisation and remote triggers. For example, Canon point-and-shoot cameras can be modified with the Canon Hacker Development Kit (CHDK, http://chdk.wikia.com/) firmware to allow the camera to be triggered from the camera's USB port with either a wired or wireless trigger. In comparison, DSLR cameras offer superior image quality but also have the disadvantages of increased cost and bulk. Most DSLRs often do not need modification for remote triggering, and the major brands have several different options for wired and wireless triggering. Another consideration when choosing cameras is the required temporal resolution of the data that need to be collected. Point-and-shoot cameras have a slower recycle time, the rate at which they can record images, which is usually 2–5 seconds between images. DSLRs recycle significantly faster and can have speeds of 1–10 frames per second.

In recent experiments, the best geometries for ISfM reconstructions have been convergent views from multiple elevations. The multiple elevations provide more parallax between cameras, which has resulted in better models. Considering ground control, the best ISfM results have been obtained from in-scene markers surveyed with a TS.

There have been three main pitfalls: synchronisation, camera clock drift, and processing. Synchronising the shutters is paramount to ensure that the cameras are all capturing the same instant. Synchronising the cameras to a hundredth of a second is easily obtainable and is sufficient for most applications. To make sorting and processing more efficient, it is best to use the image time stamps. Unfortunately, the time-keeping circuits in most consumer digital cameras have a tendency to drift, meaning that the time stamps for image sets may not match. To solve this problem you can manually adjust the time stamps to account for any clock drift using exchangeable image file format (EXIF) tag editor like ExifTool (Harvey 2014) that can batch process images. The volume of data produced by ISfM requires some of the processing and analysis to be scripted. Several of the SfM software packages have batch processing and scripting capabilities that can streamline the processing, and the analysis of the outputs can be scripted in a wide range of data analysis or GIS software.

### Main findings

This case study presents two examples from experiments with ISfM: river water surface and lava flow mapping. Using ISfM to map water surfaces in rivers can give researchers more accurate water surface elevations and slopes as inputs into hydrologic models and could provide insight into how surface roughness relates to bed roughness and other elements of hydraulics. The test was conducted in a sand bedded section of Cape Creek at the Heceta Head State Scenic Area in Oregon, and the images were collected with five Canon Powershot A3300 cameras. The ISfM model (Fig. B7.2i) shows the standing wave in the middle of the stream, and subsequent models show the wave migrating upstream as part of an antidune complex, indicating that this small section of the stream contained supercritical flow.

The second example was part of physical lava experiments at the Syracuse University Lava Project investigating the interaction of molten basalt with obstacles (Fig. B7.2ii). This research used the same 10 Canon Powershot A3300 cameras mounted 2.5 m above the flow. The 3D data from these experiments were useful in monitoring the flow characteristics, flow thickness, volumetric flux, and interaction with the obstacle through time. With four TS-surveyed control points, the RMSE of the models was 7 mm with an average point density of 750,000 points per square metre.

(a)

(b)

**Figure B7.2i** Water surface mapping with ISfM: camera setup detail with custom wireless trigger and 3D printed mount (a); photograph of the standing wave, water surface is approximately 2 m wide (b); natural colour triangular irregular network (TIN) model of the wave from ISfM (c); and 3D model coloured by elevation with hillshade enhancement (d).

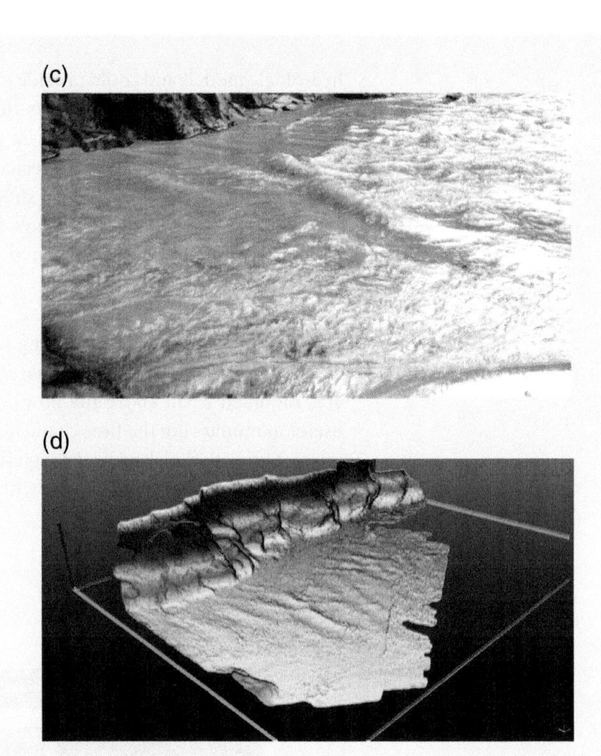


**Figure B7.2i** (*Continued*)

**Figure B7.2ii** Syracuse Lava Project experiments: experimental setup with five ISfM camera visible on the far side of the experiment platform (a); camera detail with wireless trigger and aluminium foil heat shield (b); natural colour point cloud of one of the experimental lava flows (c); and elevation coloured point cloud of the same flow (d).

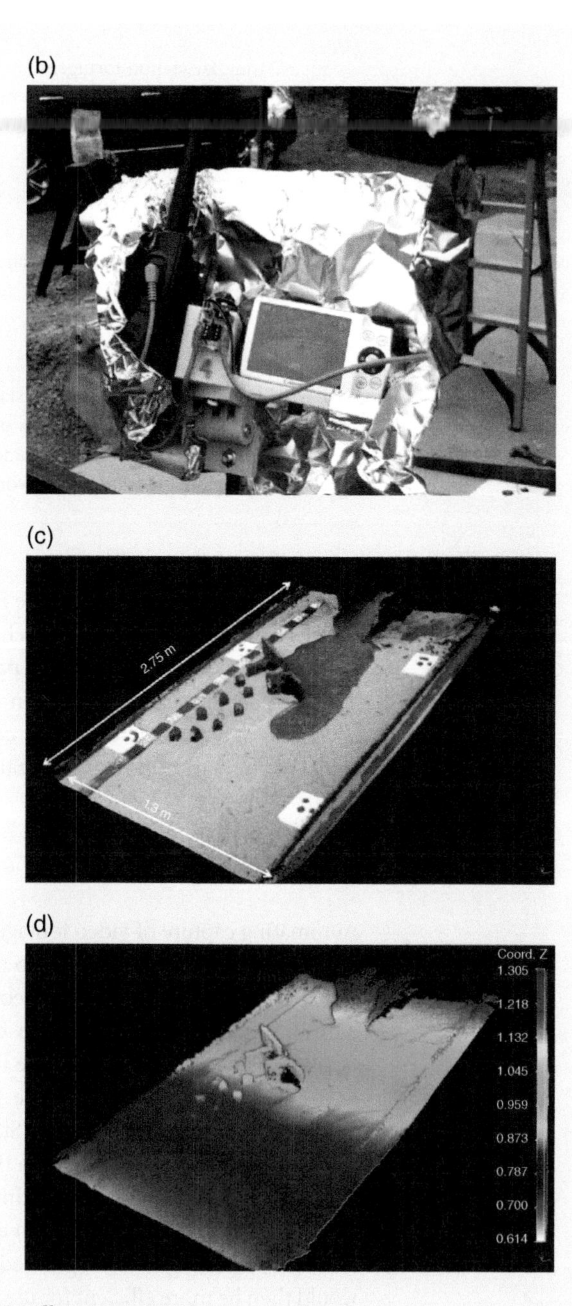


**Figure B7.2ii** (*Continued*)

### Key points for discussion

- Validation and error analysis: traditional 3D validation techniques and collecting additional checkpoints are not possible on moving surfaces. Therefore, error/uncertainty analysis may be limited to the static portions of the scene or may rely solely on the error metrics of the GCPs.

- 3D time-lapse monitoring: For both short- and long-term time-lapse projects, replacing traditional one-camera time-lapse setups with ISfM would create robust 3D data sets that could quantify the temporal change in the scene.

### Summary

ISfM has the potential to be a very powerful tool in mapping and monitoring dynamic landforms, both in the laboratory and in the field for a wide range of geoscience applications.

### Acknowledgements

I would like to thank Mark Fonstad (University of Oregon) for funding the development of the first iterations of ISfM. I would also like to thank Hannah Dietterich (University of Oregon) and Kathy Cashman (University of Bristol) for collaborating on and funding the work at the Syracuse Lava Project.

difference was noted in successive images using basic image processing, then the entire array was triggered. Such "automatic detection" is quite well developed outside of the geosciences, particularly in the automotive industry for assisted parking and for collision prevention systems, for example. In the geosciences, automatic detection could be used to issue a presence/absence signal as an "alarm" in natural hazards early warning system.

## *7.3.2 Automated Video Capture*

Automating capture of video footage to be used for SfM-MVS presents its own set of challenges in comparison to acquiring and suing still images. The most important difference is that video footage is often "shaky." Both external hardware solutions and software code solutions presently exist to remove "shake" or "blur," but in the future these might be incorporated explicitly into SfM-MVS software. One example (of several) hardware solutions is that by Ovation (2015), who produce the "StableEyes video stabiliser." This is a hardware that is simply installed in-line with the viewing monitor and is compatible with either live or pre-recorded video. In overview, the Ovation system operates by employing ultra-fast video motion analysis techniques to remove the shake in real time, resulting in co-registered frames, or stable video (Fig. 7.2), which would then be more efficient for use in an SfM-MVS workflow.

One example (of several) software code for video stabilisation is that to produce "hyper-lapse videos" by Kopf et al. (2014, 2015). They have been particularly motivated by camera shake in first-person videos where simple frame sub-sampling coupled with existing video stabilisation methods does not work because the erratic camera shake is amplified by the speed-up in camera motion. Their published algorithms firstly reconstruct the 3D input camera path as well as dense, per-frame 3D point geometry (Fig. 7.3a).

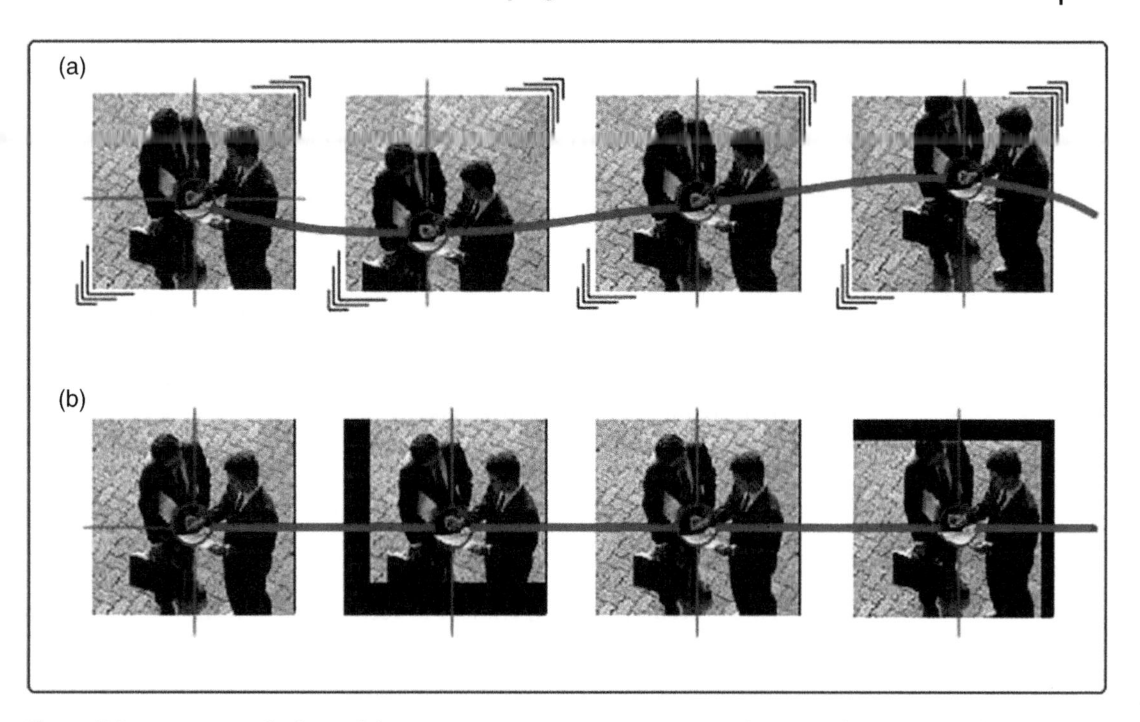


**Figure 7.2** Impression of video-stabiliser operation as with Ovation StableEyes, where an initial sequence of unstabilised images (a) are co-registered (in real time) to remove translations and rotations, producing a stabilised sequence (b). Source: http://www.ovation.co.uk/video-stabilization.html.

**Figure 7.3** Hyper-lapse video creation of a rock face whilst scrambling, that is, with extremely destabilised platform and resultant "shaky" images, by processing stages of 3D camera and 3D point cloud recovery, followed by smooth path planning for scene reconstruction (a), estimation of 3D point per camera (b) and source frame selection, seam selection, and Poisson blending for image stitching (c). Source: Kopf, J., Cohen, M.F & Szeliski, R. (2015) First-person hyper-lapse videos. http://research.microsoft.com/en-us/um/redmond/projects/hyperlapse/ [accessed on January 2015].

They then optimise a novel camera path for the output video that is smooth and passes near the input cameras whilst ensuring that the virtual camera looks in directions that can be rendered well from the input. Geometric properties *for each input frame* – that is, "proxy geometry" – are then computed (Fig. 7.3b), permitting the rendering of the frames from the novel viewpoints on the optimised path. Finally, Kopf et al. (2014, 2015) generate the novel smoothed, time-lapse video by rendering, stitching, and blending appropriately selected source frames for each output frame (Fig. 7.3c). In comparison to previous approaches, that of Kopf et al. (2014, 2015) reconstructs a full 3D camera path and world model. This enables smoothing the camera path in space-time and generating an output video with a constant-speed camera. Just as importantly and arguably most impressively, the Kopf et al. (2014, 2015) method can fill the missing regions in the video by stitching together pixels from multiple input frames. Examples by Kopf et al. to demonstrate this method of coping with extreme camera "shake" include reconstructing rock faces whilst scrambling/climbing and walking/biking with a helmet-mounted camera: http://research.microsoft.com/en-us/um/redmond/projects/hyperlapse/

In a laboratory setting, or perhaps even in a relatively controlled field setting, a ground-based video camera could be mounted on a track-based trolley for imaging experiments. For example, this setup would be well suited to a laboratory flume with sand bars evolving due to changing flow regime or base-level changes, and at pre-determined intervals a video camera "orbiting" the experiment to record images from which to derive planimetric maps and DEMs of the sand bar changes.

### *7.3.3 Crowd-Sourced Images*

Crowd sourcing of images via the internet, termed "photo tourism," was put forward by Snavely et al. (2008) and arguably is what drove early SfM-MVS development. However, whilst crowd sourcing is used in other disciplines, such as for dense urban 3D models by Irschara et al. (2012), and in geoarchaeology via the "million image database" project. Geoscience applications of SfM-MVS have not yet exploited this "remote remote-sensing" opportunity, either from images or videos.

This lack of crowd sourcing is most likely due to geoscience applications: (i) wishing to quantify error and uncertainty and thus requiring a georeferenced point cloud, (ii) an inability to search for images using a geographical position instead of a keyword, or (iii) an inability to be sure of when the images were taken and the risk of acquiring a 3D point cloud of a dynamic scene that is the amalgamation of several periods of time. Furthermore, crowd-sourced data often have limited viewpoints: Flickr images are frequently clustered around landmarks, and street view images are at present largely limited to roads (although are increasingly expanding to include other landforms, e.g. see street view/treks). As a result, these data sources can have coverage gaps in areas that have not yet been densely imaged, leaving much to be desired (Sweeney et al. 2013).

Practically, crowd sourcing has several major benefits over manual data acquisition including cost, speed, and coverage. Furthermore, a participatory approach to science must be seen to be mutually beneficial to all involved. The rapid growth of online photo collections has allowed for an efficient 3D reconstruction of massive (thousands of images) data sets (Snavely et al.

2006, 2008). However, once built, these massive models are treated as static and can only incorporate new data with immense effort, often requiring bundle adjustment as discussed in Section 3.5 on the entire set of 3D points and cameras. Sweeney et al. (2013) propose a solution to enable crowd sourcing a global 3D model by making full use of visual data acquired during augmentation by bringing SfM-MVS and simultaneous localization and mapping (SLAM) systems to operate cooperatively using the commonalities between them. Augmentation and SLAM are discussed further in Section 7.6.

There is therefore now capacity and capability for a "democratisation" of digital surveys as they require only a standard camera and computer. However, crowd-sourced 3D point clouds are likely to be "floating," not being in a global coordinate system, and may not even be sufficiently scaled. Depending on the intended use of the point cloud, approximate GCPs could be used from pre-existing fine resolution data sets (ALS, TLS, or even SfM-MVS), or from relatively coarse resolution data sets that are always improving. Therefore, whilst plenty of 3D point clouds could become available via crowd sourcing, a major challenge would remain if geoscience were to try and make use of these: that challenge being the *accurate* georeferencing of these point clouds (Section 2.3.2). However, once a georeferenced and scaled survey has been conducted, repeat surveys would not necessarily require additional data, providing sufficient number of keypoints can be identified between each successive 3D model.

Conceptually, SfM-MVS using crowd-sourced images has the capability to create 3D landforms at a spatial resolution far finer than any global DEM or indeed any aerial photogrammetry. At an overview level then crowd-sourced images will be useful for the geosciences in terms of 3D visualisation of remote landforms. For research, and by way of example, we might consider that there is a good opportunity for crowd-sourcing images of natural hazard phenomena effects, such as flooding-induced landform and infrastructure changes/damage, because (i) people tend to take images of them, (ii) relatively large errors may be permissible, and (iii) distributed data (or indeed any data) can be hard to come by through standard monitoring techniques due to the infrequency of natural hazards events and the tendency of those events to destroy *in situ* monitoring equipment.

It is likely that the uncertainty in crowd-sourced 3D models will be far less than the spatial resolution of otherwise available (global or local) elevation models. Therefore, with some caution as to identifying when the images were acquired, crowd-sourced 3D point clouds and DEMs have great potential for the geosciences as baseline data sets, from which surface changes of the order of a metre or so should be possible to identify confidently. Indeed the recent book by Nicosevici and Garcia (2013) has provided an accurate and efficient solution to large-scale scene modelling via a novel SfM-MVS algorithm that increases mapping accuracy by registering camera views directly with maps, specifically detection of images corresponding to the same scene region, or "crossovers" are used in conjunction with global alignment methods to highly reduce estimation errors, especially when mapping large areas. We also note that several consumer-grade cameras now have embedded GPS accurate to only approximately meter scale, but they could be used as initial camera positions to be used iteratively through SfM-MVS; the error can be stated in most software.

For further consideration of using crowd-sourced images, Boulos et al. (2011) gave an in-depth review of the key issues and trends in these areas and the challenges faced when reasoning and making decisions with real-time crowd-sourced data (e.g. issues of information overload, "noise," misinformation, and bias and trust).

## 7.4 Efficient Management and Manipulation of Photographs

An ongoing challenge for SfM-MVS-related work is simply managing the volume of data. As Hardin and Jensen (2011) noted, hundreds of photographs can be acquired in just a few minutes. Therefore, for geoscience applications of SfM-MVS to expand, there could usefully be developments made in (i) image browsing, organising, and mosaicking; (ii) segmentation and recognition; and (iii) object/camera positioning. In both cases, perhaps the most efficient and interesting way forward would be for geoscientists to collaborate with computer programmers and develop software using a crowd-sourcing approach. Crowd-sourcing software development (note very different to crowd-sourcing images or point clouds) is an emerging area of software engineering and is an open call for participation in any task of software development, including documentation, designing, coding, and testing. Platforms exist for managing the crowd-sourcing software development processes – a popular one is "TopCoder" (2015).

### *7.4.1 Image Browsing and Organisation*

Advanced image browsing can be permitted via the application of tags and metadata. For example, tags that can be employed to filter within a Google image search include size, colour, type, layout, people, date, and licence. Some advanced image browsers even offer automatic analysis of image content, such as face recognition and geolocation. However, most image browsers only support one categorisation at a time (e.g. folder name or time stamp), do not support relationships between tags, and do not facilitate grouping based on content. Browsing images manually (visually) can be tedious. Consequently, if SfM-MVS for the geosciences is to expand in its usage of crowd-sourced images, there is a need to go beyond current image metadata, at least to include geo-location and preferably global coordinates, but also preferably by taking into account the image content itself to perform integrated browsing based on visual characteristics as well as keywords.

To address the problem of search speed and to enable grouping of images based on content, Edmundson and Schaefer (2013) have presented a method based around an interactive image database navigation application using the Huffman table inherent in JPEG files as features, and from principal component analysis of this table they project image thumbnails onto a 2D visualisation space. Images are dynamically placed into a grid structure and organised in a tree-like hierarchy for interactive browsing of photo search results. Most impressively, Sigurþórsson et al. (2013) have presented *PhotoCube*, which has a graphical user interface (GUI) supporting three browsing dimensions which can be used simultaneously (Fig. 7.4), a multi-dimensional data model that permits applications of sets of filters, and a series of plug-ins used to define actions at different levels of enquiry, for example.

## 7.5 Point Cloud Generation and Decimation

SfM-MVS workflows are readily capable of providing large data sets in the order of a billion points: examples exist of thousands of photographs being used to generate models (e.g. main University of Maryland, Baltimore County (UMBC) Aerial Ecosynth), but these large data sets take a lot of computing power to produce and quickly become difficult to process and work with. Improvements in technologies that optimally reduce image resolution and optimally adjust image quality and improvements in technologies that seek to reconstruct images between successive views or frames may help to reduce the computational time for image processing using an SfM-MVS workflow. It is likely that graphics processing units (GPUs) and "cloud-accelerated" processing will be exploited in the future for large SfM-MVS data sets.

There is often a need to filter (see Section 4.7) or more simply to reduce or "decimate" 3D point clouds to unified 2D rasters or DEMs (see Section 4.8). However, this decimation often represents a substantial loss of data. In particular, studies concerned with sub-vertical surfaces (i.e. rock faces; see Fig. 4.10, or overhangs in gullies; Castillo et al. 2012) will experience a great loss of data using planimetric grids like this, so decimation in unstructured 3D space must be considered. It is therefore useful to consider the study by Morales et al. (2011) who used a radial basis function surface statistic to smooth point data sets according to local surface features. More recently, and perhaps also more usefully for the geosciences, Lai et al. (2014) have used Epsilon-nets, which in computational geometry relate to the approximation of set of 3D points by a collection of simpler subsets. Specifically, they specified a number of 3D points, and these were then meshed in 3D by a Poisson surface reconstruction algorithm to capture rock face topology with high fidelity and with some ability to accommodate "hiding," "shading," or "occlusion" effects.

Since 3D point cloud data typically contain significant redundant information, such as the representation of planar surfaces with hundreds of

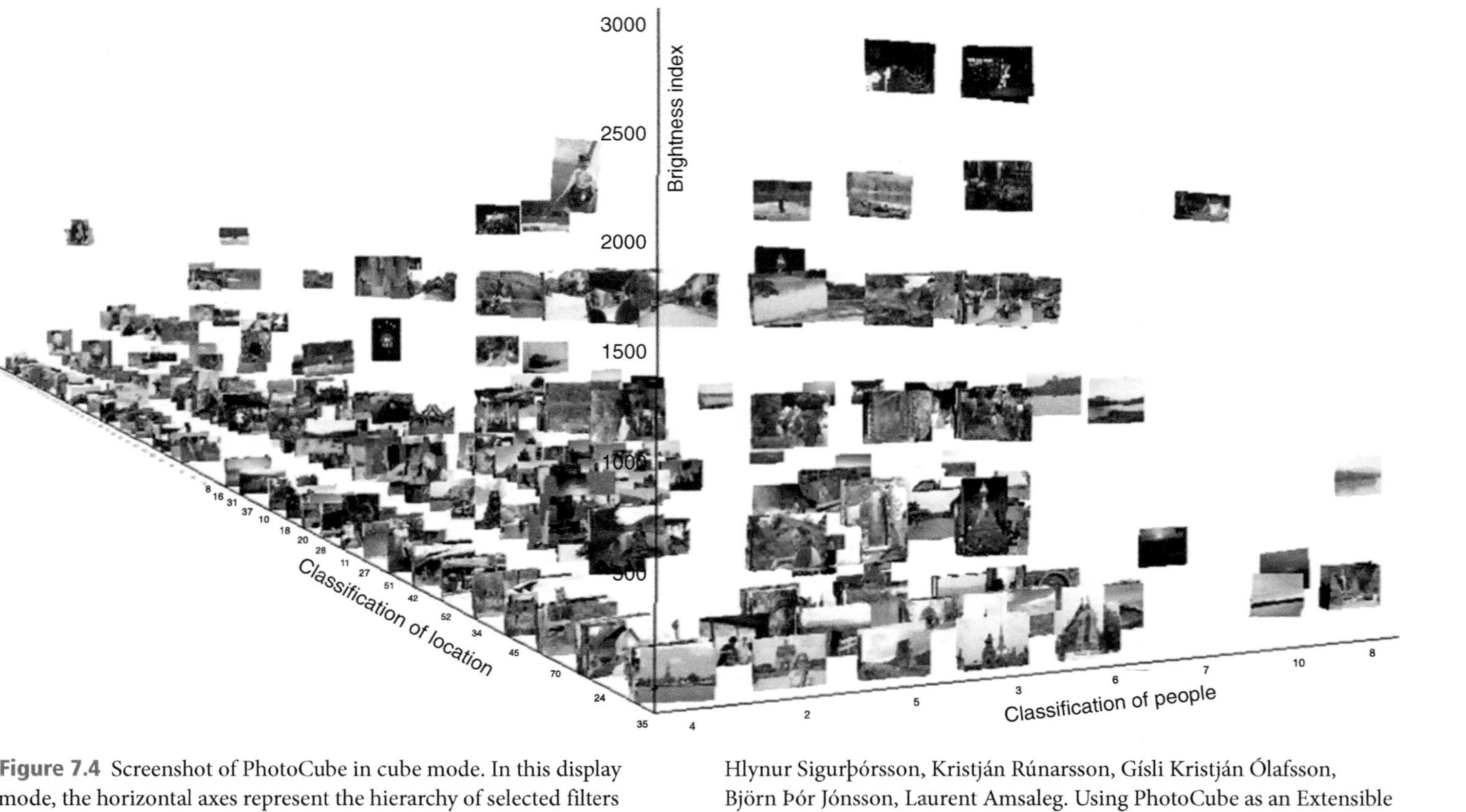


**Figure 7.4** Screenshot of PhotoCube in cube mode. In this display mode, the horizontal axes represent the hierarchy of selected filters applied to search images, and in this figure those are firstly "people" and secondly "location," each refined to "friends" and "Europe," respectively. The third axis is image brightness, as selected for considering image quality for further processing. Source: Grímur Tómasson, Hlynur Sigurþórsson, Kristján Rúnarsson, Gísli Kristján Ólafsson, Björn Þór Jónsson, Laurent Amsaleg. Using PhotoCube as an Extensible Demonstration Platform for Advanced Image Analysis Techniques. In Tenth International Workshop on Content-Based Multimedia Indexing (CBMI), Annecy, France, June, 2012.

thousands of points, techniques should be developed in the geosciences to reduce inherent redundancy. Ma et al. (2013) have presented a method for efficient triangulation, surfacing, and rendering of planar surfaces in large point clouds, and promisingly for the geosciences despite a large reduction in vertex count, the principal geometric features of each segment were well preserved. Additionally, Ma et al.'s (2013) texture generation algorithm preserves all colour information contained within planar segments, resulting in a visually appealing and geometrically accurate simplified representation.

Mapping and classification of point cloud attributes should benefit in the future from using artificial intelligence approaches, which to date have been used with laser scan data (for geological outcrop fracture analysis; Hodgetts 2013) but are directly applicable to SfM-MVS-derived point clouds. Semi-automatic methods such as neural networks, fuzzy logic, and evolutionary algorithms permit automatic classification of point-cloud data, for example, aiding the identification of varying stratigraphy or the extraction and upscaling of fault and fracture populations, especially when combined with field observations (see Hodgetts (2013) and references within).

## 7.6 Real-Time SfM-MVS and Instant Maps: Simultaneous Localisation and Mapping

Outside of the geosciences, online real-scale "mapping" has been achieved with SLAM frameworks (e.g. Dissanayake et al. 2001; Bailey and Durrant-Whyte 2006; Davison et al. 2007). SLAM is of interest to the geosciences because of the mapping element, the fact that this mapping is performed in real time, and the fact that SLAM is now being integrated with SfM-MVS. SLAM has developed in the robotics industry and specifically is the computational problem of constructing or updating a map of an unknown environment whilst simultaneously keeping track of an object within it. Applications of SLAM include self-driving cars, UAVs, autonomous underwater vehicles, planetary rovers, newly emerging domestic robots, and even relatively non-invasive human surgery. An example of SLAM becoming practical (computationally) and increasing in popularity is the recently commercialised system of Google's Project Tango: https://www.google.com/atap/project-tango/, which combines 3D motion tracking with depth sensing so a mobile device (which for SfM-MVS purposes is the camera) knows where it is and how it has moved. An example of SLAM algorithms being adopted for point-cloud data analysis is the code "6D SLAM," which enables automatic high-accurate registration of point clouds, and that is available as part of the 3DTK toolkit: http://slam6d.sourceforge.net/.

In overview, SLAM is a series of algorithms that are tailored to available computational resources. Thus to date they are generally not aimed at the perfection of 3D landform model; they create topology rather than geometry. SLAM algorithms are probability based and most commonly are either approximated by statistical techniques (either Kalman filters or Monte Carlo methods) or set membership techniques. The statistical techniques offer a more accurate estimate via a probability density function for the pose of the object and for the parameters of the map. The set membership techniques can better accommodate non-linearity and have less assumptions (i.e. independence) and estimate a set, which may be thought of as a type of envelope enclosing the object, for the pose of the object. Increasingly, SLAM is also performed via bundle adjustment algorithms (see Section 3.5). Bundle adjustment algorithms jointly estimate object pose and environment/landform positions, thereby increasing map fidelity.

SLAM and combined SfM-MVS methods still present limitations: the trajectory and 3D point cloud are not known exactly. Indeed, all the displacements and 3D positions are relative, and it is not possible to obtain an absolute localisation of each reconstructed element. Besides, in addition to being prone to numerical error accumulation, SLAM algorithms may present scale factor drift: their reconstructions are done up to a scale factor, theoretically constant on the whole sequence, but often fluctuating in practice. One solution is to use SLAM simultaneously with a coarse GIS-based model (Lothe et al. 2009) whereby "globalisation" propagates new visual information back to the model; specifically, continuous updating of 3D models is made with visual data from live users, thereby providing data to fill coverage gaps that are common in 3D reconstructions and to provide the most current view of an environment as it changes over time. More widely, Sweeney et al. (2013) considered that tracking and mapping for large-scale reconstructions that enables SfM-MVS and SLAM to operate cooperatively is a crucial step towards enabling users to augment (urban) environments with location-specific information at any location in the world for a truly global augmented reality (Section 7.7).

## 7.7 Augmented Reality

In the future the geosciences could consider augmented reality environments as an alternative to proprietary image handling/management software, 3D point cloud software, and also GIS software. Adjusting and rendering images and interacting with 3D point clouds may well prove easier in an augmented reality environment than in separate graphics software, image analysis software, and within a GIS, for example.

Augmented reality manipulation functions can include the removal or addition of features or objects in images. Where images intended for SfM-MVS include static or mobile foreground objects, such as trees or people, respectively, that permanently or temporarily obscure the landform of interest, it may be desirable to remove them from images prior to SfM-MVS point cloud generation. "PatchMatch" is the codename for Adobe Photoshop's new Content-Aware Fill feature, which enables exceptionally easy and fast editing of images including image completion or "inpainting." Inpainting does not use a second image, but rather comprises the deletion of a foreground object to create a vacant patch and then automatic infilling using patterns of the surrounding background. A user can define specific parts of the image to work on to increase the speed of computation but most impressively can define specific features that have a pattern, such as being linearly extensive, desirable for the vacant patch.

## 7.8 Detection of Object or Surface Motion: Non-Rigid SfM

Surface motion is perhaps ultimately what geoscientists seek to understand when considering form/process relationships. Where surfaces change relatively slowly, such as during soil creep and solifluction, survey costs are often too high because of the duration of observations required. Where surfaces change extremely quickly, such as lava flows, ice and rock falls, conventional topographic survey technology is not fast enough and the surface of interest is too hazardous to measure directly. The most straightforward approach to capturing motion is presently via the use of motion sensors. Although motion sensors are able to measure landform motion directly, they are intrusive. That is partly why the geosciences have seized upon SfM-MVS via fixed arrays of multiple time-lapse time-synchronised cameras to address the moving surface problem (see Boxes 7.1 and 7.2).

Outside of the geosciences, several applications of SfM-MVS to motion capture have appeared in the past few years (Hasler et al. 2009; Shiratori et al. 2011). Hasler et al. (2009) presented an approach for markerless motion capture by recording articulated objects using several unsynchronised moving cameras. In this system, the reconstruction of static landform and camera registration was performed using the SfM-MVS method, based on which both the positions and the joint configurations of subjects were able to be recovered.

Where a landform changes *during* a survey, SfM-MVS assumptions are invalid – most notably that the scene is rigid and thus that apparent landform changes are due solely to the camera motion. There is therefore "an explosion of unknowns" (Sheikh & Khan 2010) with measuring a moving landform because every point on that landform, which in rigid SfM-MVS has a single $x$, $y$, $z$ coordinate, has multiple coordinates (i.e. one per time step). However, the motion of a landform is not random: 3D points are often highly correlated in space and time because they move due to an applied force, and hence their acceleration is limited by the force and, therefore, the landform does not change arbitrarily over time.

Furthermore, 4D structure often lies in a low-dimensional sub-space, that is, topology. Therefore, the two prevailing approaches to solving "non-rigid SfM" (NRSfM) are based on shape and trajectory (Fig. 7.5; Table 7.1).

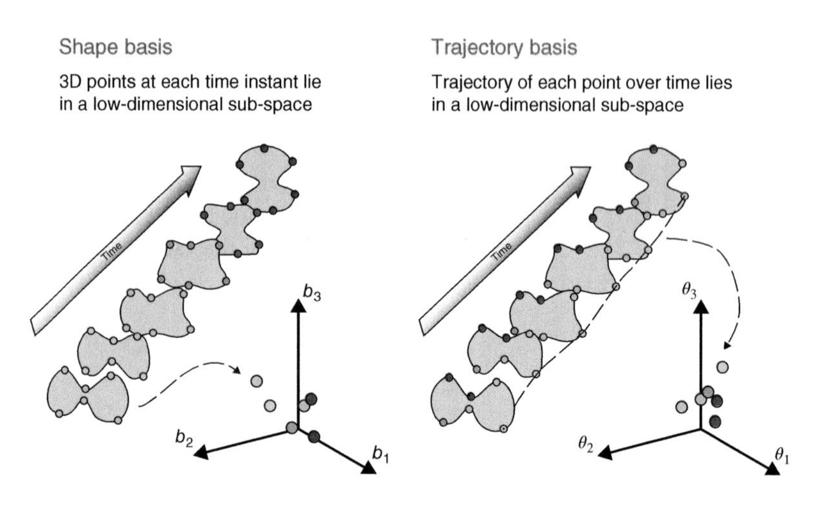


**Figure 7.5** Two major approaches to NRSfM. Note that for clarity the deforming structure depicted is a planar membrane, but in reality it will be a deforming 3D structure. Source: Sheikh, Y. & Khan, S. (2010) *Non-rigid SfM tutorial slides.* http://www.cs.cmu.edu/~yaser/ECCV2010Tutorial.html [accessed on January 2015].

**Table 7.1** Pros and cons of approaches to NRSfM.

| | **Shape** | **Trajectory** |
|---|---|---|
| Model | Can be learnt | Hard to specialise |
| Specificity | Object dependent | Generalised |
| Frame ordering | Irrelevant | Exploited |
| Point ordering | Exploited | Irrelevant |

Source: After Sheikh, Y. & Khan, S. (2010) *Non-rigid SfM tutorial slides.* http://www.cs.cmu.edu/~yaser/ECCV2010Tutorial.html [accessed on January 2015].

In overview, and in comparison to rigid SfM-MVS as described fully in Chapter 3, the trajectory approach according to Sheikh and Khan (2010) firstly takes point correspondences and EXIF data to estimate the camera matrices using a RANSAC framework (see Section 3.4). Secondly, because triangulation-based solutions are not applicable as multiple views of the point may not exist at each instant in time, it then uses the camera matrices and dynamic point correspondences to create overloaded linear system using discrete cosine transform (DCT). This transform is a signal data and image compression method (e.g. it is used in JPEG and MP3 formats) identifying that almost all signal "energy" is concentrated in the low-frequency area. Using DCT, a discrete data series is reduced in terms of cosine functions oscillating at discrete frequencies. Finally, the trajectory approach then performs a bundle adjustment (see Section 3.5).

A key consideration of the trajectory approach is definition of the trajectory basis, which is a reduction of the trajectory to low-dimensional subspace (Park et al. 2010), and specifically how to tune the basis size and how to choose the best basis combination (Valmadre & Lucey 2012). Wang et al. (2014, 2015) presented an automatic method to select the trajectory basis, which can select appropriate basis size because if the basis size is too small, the trajectory is poorly represented by the basis, but too large basis size makes a system more ill conditioned and the reconstruction error becomes unbounded. Thus Wang et al. (2014, 2015) claim improved reconstruction accuracy and efficiency of the trajectory approach to NRSfM.

Both shape and trajectory approaches to NRSfM seem to work in simple cases as presented by Sheikh and Khan (2010), that is, tens of points over hundreds of time steps, but lead those authors to conclude that perhaps it is necessary to de-correlate camera motion and object motion, which could perhaps be achieved by simultaneously imaging a moving landform from multiple positions, that is, using camera arrays as in Boxes 7.1 and 7.2).

The main advances in NRSfM research are practically limited in deploying spatiotemporal bilinear models to highly under-constrained 3D motion data. In other words, there are two big practical problems with NRSfM. Firstly, NRSFM data can be missing during measurement due to projection, occlusion, or miscorrespondence. Missing data issues are present in "normal" or rigid SFM-MVS but are of greater significance in dynamic 3D reconstruction because the observation system has only one opportunity to directly measure information about the structure at a particular time instant. Thus, the question at the core of NRSfM is what internal model a system should refer to when there is insufficient information; ideally, a good model should capture all spatial, temporal, and spatiotemporal correlations in the data as these correlations permit reasoning about the information that is missing.

Secondly, NRSfM data are voluminous, to the point of almost being unmanageable, for example, 100 points over 120 time steps is 36,000 degrees of freedom, and therefore the number of possible correlations is approximately 648 million parameters. "Learning" these correlations requires a large quantity of samples, where each sample is a full spatiotemporal sequence. For most applications, such large numbers of sequences are not accessible or practically manageable. In answer to these problems of missing data and voluminous data, Simon et al. (2014) present a probabilistic model of 3D data that captures most salient correlations and can still be estimated from a few or even one sequence.

The most obvious applications of NRSfM in the geosciences will be where surfaces change extremely quickly and/or are too hazardous to directly measure. Applications will include laboratory and field studies. Phenomena could include landslides, lava and debris flows, and ice and rock falls.

## 7.9 Summary

This chapter has considered the likely future developments in hardware, platforms, and cameras that are desired to facilitate further development of SfM-MVS in the geosciences. Progressive automation of acquisition will be enabled by the usage of multiple cameras simultaneously, by automated video capture, by cameras triggered by an environmental sensor, and by crowd-sourced images. The first three of these could be used in combination. Automated video capture could benefit from a number of image-stabilisation products (hardware and software) now available. Efficient management and manipulation of photographs can be progressed with developments in image browsing, organisation, mosaicking, segmentation, and recognition. Improvements in technologies that optimally reduce image resolution and optimally adjust image quality and improvements in technologies that seek to reconstruct images between successive views or frames may help to reduce the computational time for image processing using an SfM-MVS workflow.

The geosciences could usefully utilise real-time mapping or SLAM technology. For example, for real-time detection of landforms, classification of landforms and quantification of changes would all make interpretations of environmental processes much more objective and hence would make understanding more immediate and more complete. SLAM also has utility in point-cloud processing, for example, automatic highly accurate registration of point clouds.

The geosciences could explore augmented reality environments as an alternative to proprietary image processing software, point cloud software, and GIS. Rendering and interacting with raw images, for example, to remove foreground objects masking a view of a landform of interest, and manipulating 3D point clouds may well prove easier in an augmented reality than in separate graphics software or within a GIS.

The potential for NRSfM methods to be adapted and accommodated by the geosciences needs further exploration, but there is a precedent of NRSfM in other disciplines, and arguably NRSfM is the biggest as-yet-unrealised and most exciting development that could occur for SfM in the geosciences.

## References

Bailey, T. & Durrant-Whyte, H. (2006) Simultaneous localization and mapping (SLAM): part II. *IEEE Robotics and Automation Magazine*, **13** (**3**), 108–117.

iBeacon (2015) http://www.ibeacon.com/. [accessed on January 2015].

Bitelli, G., Dubbini, M. & Zanutta, A. (2004) Terrestrial laser scanning and digital photogrammetry techniques to monitor landslide bodies. *International Archives of Photogrammetry, Remote Sensing and Spatial Information Sciences*, **35**, 246–251.

Boulos, M.N.K., Resch, B., Crowley, D.N. et al. (2011) Crowdsourcing, citizen sensing and sensor web technologies for public and environmental health surveillance and crisis management: trends, OGC standards and application examples. *International Journal of Health Geographics*, **10** (**1**), 67.

Castillo, C., Pérez, R., James, M.R., Quinton, J.N., Taguas, E.V. & Gómez, J.A. (2012) Comparing the accuracy of several field methods for measuring gully erosion. *Soil Science Society of America Journal*, **76** (**4**), 1319–1332.

Chandler, J.H., Wackrow, R., Sun, X., Shiono, K. & Rameshwaran, P. (2008) Measuring a dynamic and flooding river surface by close range digital photogrammetry.

*Presented at the Silk Road for Information from Imagery*, pp. 211–216. Beijing. http://www.isprs.org/proceedings/XXXVII/congress/8_pdf/2_WG-VIII-2/09.pdf [accessed on 2 November 2014]

Colomina, I. & Molina, P. (2014) Unmanned aerial systems for photogrammetry and remote sensing: a review. *ISPRS Journal of Photogrammetry and Remote Sensing*, **92**, 79–97.

Davison, A.J., Reid, I.D., Molton, N.D. & Stasse, O. (2007) MonoSLAM: real-time single camera SLAM. *IEEE Transactions on Pattern Analysis and Machine Intelligence*, **29** (**6**), 1052–1067.

Dissanayake, M.G., Newman, P., Clark, S., Durrant-Whyte, H.F. & Csorba, M. (2001) A solution to the simultaneous localization and map building (SLAM) problem. *IEEE Transactions on Robotics and Automation*, **17** (**3**), 229–241.

Edmundson, D. & Schaefer, G. (August 2013) A browsing approach to online photo search results. In: *2013 IEEE International Conference on Signal Processing, Communication and Computing (ICSPCC)*, August 5–8, 2013, pp. 1–5. IEEE, Yunnan, China.

Fonstad, M.A., Dietrich, J.T., Courville, B.C., Jensen, J.L. & Carbonneau, P.E. (2013) Topographic structure from motion: a new development in photogrammetric measurement. *Earth Surface Processes and Landforms*, **38** (**4**), 421–430.

Hardin, P.J. & Hardin, T.J. (2010) Small-scale remotely piloted vehicles in environmental research. *Geography Compass*, **4** (**9**), 1297–1311.

Hardin, P.J. & Jensen, R.R. (2011) Small-scale unmanned aerial vehicles in environmental remote sensing: challenges and opportunities. *GIScience & Remote Sensing*, **48** (**1**), 99–111.

Harvey P. (2014) ExifTool. http://www.sno.phy.queensu.ca/~phil/exiftool/ [accessed on January 18, 2016].

Hasler, N., Rosenhahn, B., Thormahlen, T., Wand, M., Gall, J., & Seidel, H. P. (2009, June) Markerless motion capture with unsynchronized moving cameras. In: *IEEE Conference on Computer Vision and Pattern Recognition, 2009. CVPR 2009*, pp. 224–231. IEEE.

Hodgetts, D. (2013) Laser scanning and digital outcrop geology in the petroleum industry: a review. *Marine and Petroleum Geology*, **46**, 335–354.

Irschara, A., Zach, C., Klopschitz, M. & Bischof, H. (2012) Large-scale, dense city reconstruction from user-contributed photos. *Computer Vision and Image Understanding*, **116** (**1**), 2–15.

James, M.R. & Robson, S. (2014) Sequential digital elevation models of active lava flows from ground-based stereo time-lapse imagery. *ISPRS Journal of Photogrammetry and Remote Sensing*, **97**, 160–170.

Jensen, A. M., Baumann, M., & Chen, Y. (2008, July) Low-cost multispectral aerial imaging using autonomous runway-free small flying wing vehicles. In: *Geoscience and Remote Sensing Symposium, 2008. IGARSS 2008. IEEE International*, Vol. **5**, pp. V-506. IEEE.

Kopf, J., Cohen, M.F. & Szeliski, R. (2014) First-person hyper-lapse videos. *ACM Transactions on Graphics*, **33** (**4**), 78.

Kopf, J., Cohen, M.F & Szeliski, R. (2015) First-person hyper-lapse videos. http://research.microsoft.com/en-us/um/redmond/projects/hyperlapse/ [accessed on January 2015].

Lai, P., Samson, C. & Bose, P. (2014) Visual enhancement of 3D images of rock faces for fracture mapping. *International Journal of Rock Mechanics and Mining Sciences*, **72**, 325–335.

Lothe, P., Bourgeois, S., Dekeyser, F., Royer, E. & Dhome, M. (June 2009) Towards geographical referencing of monocular slam reconstruction using 3d city models: application to real-time accurate vision-based localization. In: *IEEE Conference on Computer Vision and Pattern Recognition, 2009. CVPR 2009,* June 20–25, 2009, pp. 2882–2889. IEEE, Fontainebleau Resort, Miami Beach, FL.

Ma, L., Whelan, T., Bondarev, E., de With, P.H. & McDonald, J. (September 2013) Planar simplification and texturing of dense point cloud maps. In: *2013 European Conference on Mobile Robots (ECMR),* September 25–27, 2013, pp. 164–171. IEEEBarcelona, Spain.

Morales, R., Wang, Y. & Zhang, Z. (2011) Unstructured point cloud surface denoising and decimation using distance RBF K-nearest neighbor Kernel. In: *Advances in Multimedia Information Processing-PCM 2010,* pp. 214–225. Springer Berlin Heidelberg.

Musialski, P., Wonka, P., Aliaga, D.G., Wimmer, M., Gool, L. & Purgathofer, W. (2013) A survey of urban reconstruction. *Computer Graphics Forum,* **32** (**6**), 146–177.

Nicosevici, T. & Garcia, R. (2013) *Efficient 3D Scene Modeling and Mosaicing.* Springer, Berlin.

Ovation (2015) *Video stabilisation system.* http://www.ovation.co.uk/video-stabilization.html [accessed on January 2015].

Park, H.S., Shiratori, T., Matthews, I. & Sheikh, Y.A. (2010) 3D reconstruction of a moving point from a series of 2D projections. *European Conference on Computer Vision,* **6313**, 158–171.

Sheikh, Y. & Khan, S. (2010) *Non-rigid SfM tutorial slides.* http://www.cs.cmu.edu/~yaser/ECCV2010Tutorial.html [accessed on January 2015].

Shim, M., Yilma, S., & Bonner, K. (2008). A robust real-time structure from motion for situational awareness and RSTA. In *SPIE Defense and Security Symposium* (pp. 696205-696205). International Society for Optics and Photonics.

Shiratori, T., Park, H.S., Sigal, L., Sheikh, Y. & Hodgins, J.K. (2011) Motion capture from body-mounted cameras. *ACM Transactions on Graphics,* **30** (**4**), 31.

Sigurþórsson, H., Tómasson, G., Jónsson, B.Þ. & Amsaleg, L. (2013) PhotoCube: towards multi-dimensional image browsing. http://en.ru.is/media/technical-reports/RUTR-CS12001.pdf [accessed on January 2015].

Simon, T., Valmadre, J., Matthews, I. & Sheikh, Y. (2014) Separable spatiotemporal priors for convex reconstruction of time-varying 3D point clouds. In: *Computer Vision–ECCV 2014,* pp. 204–219. Springer International Publishing.

Smolic, A., Kauff, P., Knorr, S. et al. (2011) Three-dimensional video postproduction and processing. *Proceedings of the IEEE,* **99** (**4**), 607–625.

Snavely, N., Seitz, S.M. & Szeliski, R. (2006) Photo tourism: exploring photo collections in 3d. *ACM Transactions on Graphics,* **25** (**3**), 835–846.

Snavely, N., Seitz, S.M. & Szeliski, R. (2008) Modeling the world from internet photo collections. *International Journal of Computer Vision,* **80** (**2**), 189–210.

Sweeney, C., Hollerer, T. & Turk, M. (2013) *Improved outdoor augmented reality through "Globalization".* http://ieeexplore.ieee.org/stamp/stamp.jsp?tp=&arnumber=6671820 [accessed on January 18, 2016].

TopCoder (2015) https://www.topcoder.com/ [accessed on January 2015].

Valmadre, J. & Lucey, S. (2012). General trajectory prior for non-rigid reconstruction. In: *2012 IEEE Conference on Computer Vision and Pattern Recognition (CVPR),* June 16–21, 2012, pp. 1394–1401. Providence, RI.

Wang, Y.M., Zheng, J.B., Jiang, M.F., Xiong, Y.L. & Huang, W.Q. (2014) A trajectory basis selection method for non-rigid structure from motion. *Applied Mechanics and Materials*, **644–650**, 1396–1399.

Wang, Y., Zheng, J., Jiang, M., Xiong, Y. & Huang, W. (2015) Application of automatic trajectory basis selection method for non-rigid structure from motion. *Journal of Information and Computational Science*, **12**, 503–513.

Whitehead, K. & Hugenholtz, C.H. (2014) Remote sensing of the environment with small unmanned aircraft systems (UASs), part 1: a review of progress and challenges 1. *Journal of Unmanned Vehicle Systems*, **2** (**3**), 69–85.

## Further Reading/Resources

Readers should be aware of a recent book entitled *Computer Vision Analysis of Image Motion by Variational Methods* by Mitchie and Aggarwal (2014) that has four core themes of "motion estimation," "detection," "tracking," and "three-dimensional interpretation": Mitiche, A. & Aggarwal, J.K. (2014) *Computer Vision Analysis of Image Motion by Variational Methods*. Springer, Cham.

The recently commercialised system of Google's Project Tango: https://www.google.com/atap/project-tango/ Web page includes concept guides and tutorials on their implementation of motion tracking, area learning, and depth perception, the latter which includes 3D point cloud construction and analysis.

Crum et al. (2014) provide a review of the theory and practice of non-rigid image registration: Crum, W.R., Hartkens, T. & Hill, D.L.G. (2014) Non-rigid image registration: theory and practice. http://www.birpublications.org/doi/full/10.1259/bjr/25329214 [accessed on January 18, 2016].

Readers interested in the NRSfM-MVS problem may find the latest developments posed by Wang et al. (2014) useful: Wang, Y.M., Zheng, J.B., Jiang, M.F., Xiong, Y.L. & Huang, W.Q. (2014) A trajectory basis selection method for non-rigid structure from motion. *Applied Mechanics and Materials*, **644**, 1396–1399.

# 8 Concluding Recommendations

Structure from Motion–Multi-View Stereo (SfM-MVS) has revolutionised the acquisition of topographic data in a wide range of disciplines, but has notable applications in academic and industrial branches of the geosciences, for which topographic data have long played a central role, despite the time and financial cost of acquiring such data in the past. Over the past 7 years, applications of SfM-MVS in the geosciences have multiplied and continue to increase in number and breadth dramatically. This uptake is not surprising given the ease with which topographic data can now be acquired. To conclude this book, we outline briefly **six** key recommendations for the future use of SfM-MVS in the geosciences.

## 8.1 Key Recommendation 1: Get "Under the Bonnet" of SfM-MVS to Become More Critical End Users

Many complex processes and algorithms are called upon to complete the full SfM-MVS workflow. At each step, choices must be made, for example, relating to the closeness of matches demanded in the keypoint correspondence step, or the specification of an "outlier" in the random sample consensus (RANSAC) procedure to identify geometrically consistent correspondences. At almost every step, an arbitrary parameter must be introduced to either ensure data quality or to improve computer runtime. To date, geoscience users of SfM-MVS have not engaged fully with this process, preferring instead to use "black box" generic qualitative data quality

*Structure from Motion in the Geosciences*, First Edition. Jonathan L. Carrivick, Mark W. Smith, and Duncan J. Quincey.

Companion Website: www.wiley.com/go/carrivick/structuremotiongeosciences

parameters (as implemented by many commercially available software) or even leave the default settings untouched. A much more critical application of SfM-MVS by geoscientists, especially academics, is encouraged in the future, even if this necessitates greater engagement with open-source SfM-MVS codes. To some extent, those with prior photogrammetric expertise have been vocal in rightly arguing that lessons from conventional photogrammetry need to be re-learned (e.g. relating to specifying lens models); however, this critical application should extend to all steps in the full SfM-MVS workflow, including those elements originating in computer vision. Certainly, the geoscience community should lobby for common commercial software providers to provide greater flexibility in the specification of quality control parameters, as might be anticipated in any case as the technology matures.

## 8.2 Key Recommendation 2: Get Co-ordinated to Understand the Sources and Magnitudes of Error

The flexibility of the SfM-MVS approach renders pre-determination of expected errors extremely difficult. SfM-MVS surveys can be undertaken from a range of different platforms, with different sensors, over different scales, scaled and georeferenced using different methods, using different software packages and filtered or decimated into different terrain products using a range of algorithms. The number of combinations of each of these options is extremely large, and this is before we consider the large variability of environments in which we work. Quantifying the error expected in any single pipeline is a mammoth task that has only recently been attempted in the geosciences, albeit in an uncoordinated manner. As geoscientists, the publication of standard reference data sets freely available online would improve the understanding of the optimum workflows applied to a given problem, thereby enabling a maximisation of the potential of SfM-MVS in the discipline. A coordinated approach to working through every single SfM-MVS pipeline would then be required.

Moreover, there is no agreed method of validation for SfM-MVS. The dense three-dimensional (3D) point cloud generated is rarely used directly in geoscience applications. Yet some studies validate SfM-MVS based on the individual points. Other studies first generate commonly used terrain products prior to validation, but the resulting validation will be partially a function of the algorithms used for interpolation or decimation. Certainly, greater work is needed to identify the dominant sources of error in the full SfM-MVS workflow, which must be understood by academics and industries, such as environmental consultancy, alike.

## 8.3 Key Recommendation 3: Focus on the Research Question

SfM-MVS produces data that are most similar to those derived from terrestrial laser scanning (TLS), yet at a fraction of the cost. In common with conventional photogrammetry, the final achievable data quality, in terms of accuracy, precision, and resolution, is strongly influenced by survey range. At short ranges (~10 m) data quality comparable with TLS can be obtained. At longer ranges (100–200 m) data comparable to those of airborne laser scanning (ALS) is capable of being generated with SfM-MVS. Overall, this is remarkable and greatly in excess of anything imaginable just a decade ago.

However, it is human nature to continue to push more and more from what is available. For many applications, such a challenge is merely a distraction, and the geosciences should not lose sight of why a technique is used in the first place. For example, ALS and TLS have been widely available for over a decade, and given the expanded data availability and data types, perhaps only a handful of key conceptual discipline-shaping advances have emerged. The real gains to be had, therefore, are in the insightful, innovative and even imaginative application of SfM-MVS to address the "big questions" in the geosciences, but using the most appropriate workflow for the task.

## 8.4 Key Recommendation 4: Focus Your Efforts on Data Processing

The speed of acquisition of topographic data has increased by several orders of magnitude with the introduction of ALS, TLS, and now SfM-MVS into the geoscientist's toolkit. However, advances in data processing, visualisation, and application have lagged well behind and are now a well-defined bottleneck in the overall workflow. Geoscientists are very capable of creating fully 3D dense point clouds, but they typically degrade and decimate these data into often rasterised digital elevation models as they are both experienced and most comfortable in working with such data formats. Geoscientists need to develop more direct uses of 3D point cloud data, SfM-MVS-derived orthophotographs, and more flexible software environments that can cope with voluminous data. Such efforts are ongoing, and incremental improvements are almost inevitable, but should be focussed and coordinated to make the most of the data-rich future.

## 8.5 Key Recommendation 5: Learn from Other Disciplines

SfM-MVS as applied in the geosciences emerged from a combination of conventional photogrammetric techniques and algorithms and advances in computer vision and has built upon decades of advances in each discipline. SfM-MVS is not unique in that regard. Geoscientists should continue to look beyond the geosciences for other such technologies and developments that can be applied innovatively or integrated with more "home-grown" techniques. The disciplines of imaging (medical, digital, optical), robotics, and computing (communication and sensor development and software and hardware innovation) are all complementary to the SfM-MVS workflow and have their own motivation for advancing knowledge. By embracing these technical developments, geoscientists can focus less on how to acquire data and more on what to do with it, essentially turning data into information about landform dynamics.

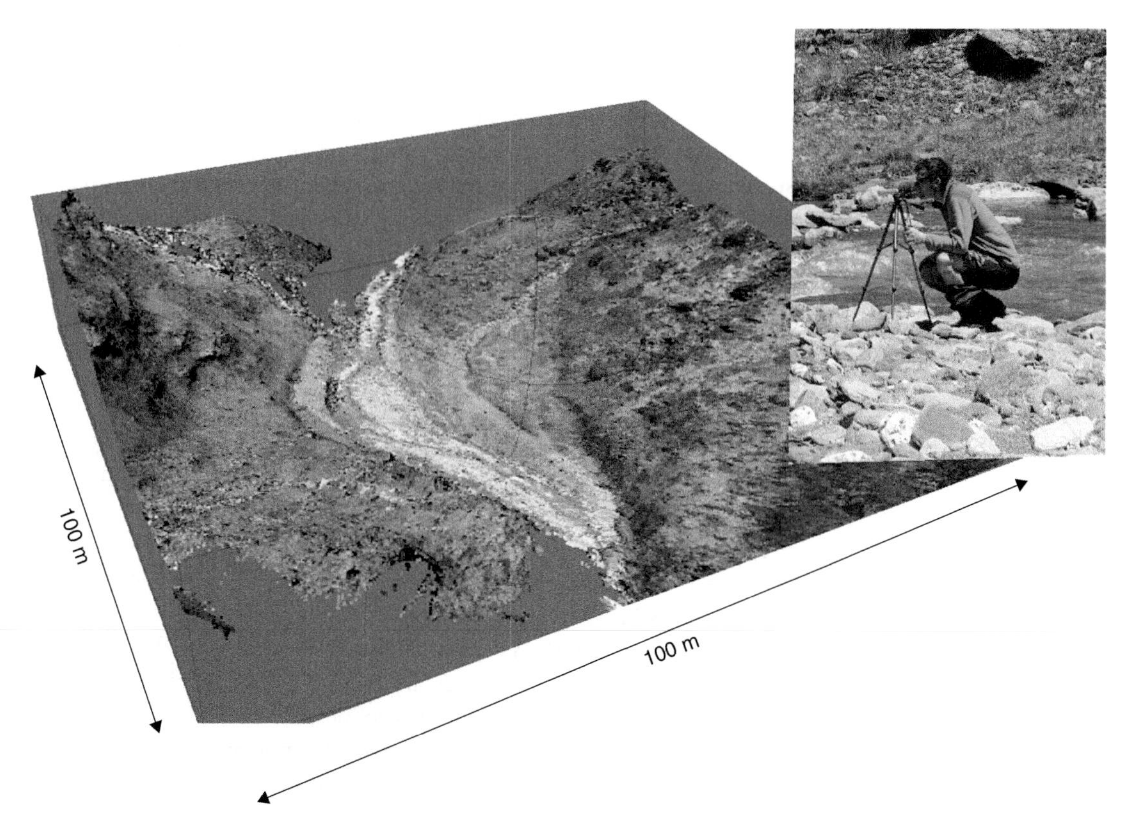


**Figure 8.1** Joseph Croisdale from the School of Geography, University of Leeds, using SfM-MVS for their dissertation concerning braided river dynamics in the upper part of the Rob Roy Valley, New Zealand.

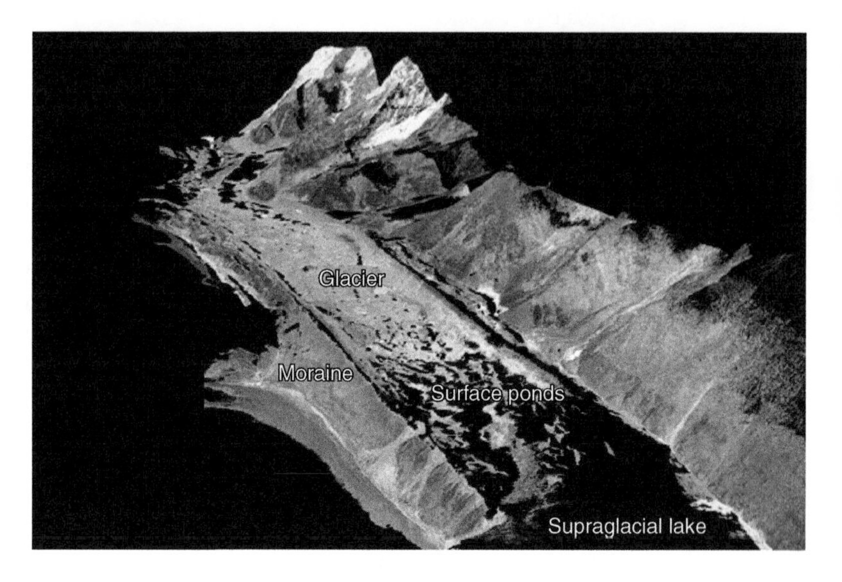


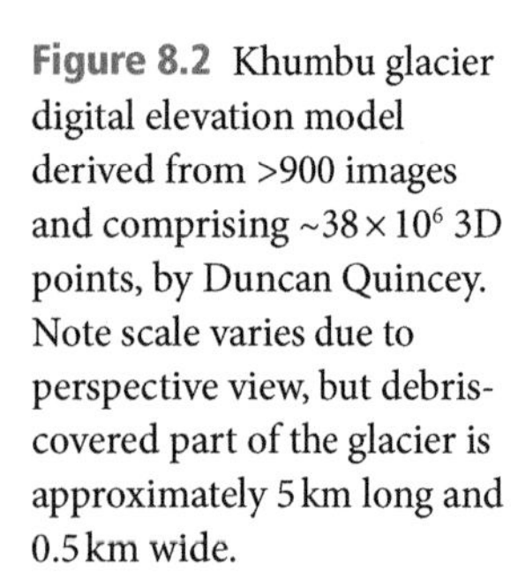

**Figure 8.2** Khumbu glacier digital elevation model derived from >900 images and comprising ~$38 \times 10^6$ 3D points, by Duncan Quincey. Note scale varies due to perspective view, but debris-covered part of the glacier is approximately 5 km long and 0.5 km wide.

## 8.6 Key Recommendation 6: Harness the Democratising Power of SfM-MVS

As often discussed, the primary advantage of SfM-MVS is its availability to anyone who owns a camera and a computer. However, to date, SfM-MVS practitioners have typically been those who are experienced in creating, manipulating, and applying dense 3D point clouds. It is to be expected that such a community would form what could be considered the "trailblazers" of SfM-MVS in the geosciences. But the geosciences must not lose sight of the potential for this technique to make a difference beyond experienced surveyors or academics. SfM-MVS is well within the grasp of undergraduates (Fig. 8.1); indeed we deliberately intended that this book should be accessible to such an audience (particularly Chapter 4 which provides practical advice on using SfM-MVS). Depending on the scale of enquiry, basic georeferencing can be provided by either a laser rangefinder or from Google Earth coordinates. Free web-based services can be used for point cloud generation. We should encourage interested undergraduate and taught postgraduate students to use SfM-MVS in their research projects – not least because we ourselves have learned much from their experiences! (Fig. 8.1).

Beyond this, the geosciences should encourage public participation in high-profile SfM-MVS-based research to foster greater public engagement with the discipline. SfM-MVS could play a central role in school outreach programs. It certainly has the necessary "wow" factor (Fig. 8.2). Whilst SfM-MVS may have originated in other disciplines, the use of SfM-MVS data is at its most visceral in the geosciences and has the potential to inspire a whole new generation of geoscientists to study the form and evolution of the Earth's surface.

# Index

(Note: Major and important occurrences of the word are in bold)



*Structure from Motion in the Geosciences*, First Edition. Jonathan L. Carrivick, Mark W. Smith, and Duncan J. Quincey.

Companion Website: www.wiley.com/go/carrivick/structuremotiongeosciences

Printed and bound by CPI Group (UK) Ltd, Croydon, CR0 4YY
16/04/2025
14658459-0005